Pathways in Mathematics

Each “Pathways in Mathematics” book offers a roadmap to a currently well developing mathematical research field and is a first-hand information and inspiration for further study, aimed both at students and researchers. It is written in an educational style, i.e., in a way that is accessible for advanced undergraduate and graduate students. It also serves as an introduction to and survey of the field for researchers who want to be quickly informed about the state of the art. The point of departure is typically a bachelor/masters level background, from which the reader is expeditiously guided to the frontiers. This is achieved by focusing on ideas and concepts underlying the development of the subject while keeping technicalities to a minimum. Each volume contains an extensive annotated bibliography as well as a discussion of open problems and future research directions as recommendations for starting new projects

More information about this series at http://www.springer.com/series/15133

Wolfgang König

The Parabolic Anderson Model

Random Walk in Random Potential

Wolfgang König
Weierstraß-Institut
für Angewandte Analysis und Stochastik
Berlin, Germany

Institute for Mathematics
TU Berlin
Berlin, Germany

ISSN 2367-3451 ISSN 2367-346X (electronic)
Pathways in Mathematics
ISBN 978-3-319-33595-7 ISBN 978-3-319-33596-4 (eBook)
DOI 10.1007/978-3-319-33596-4

Library of Congress Control Number: 2016940195

Mathematics Subject Classification (2010): 60-02, 60J55, 60F10, 60K35, 60K37, 82B44, 60J27, 60J65, 60J80, 60K40, 80A20, 82D30

Printed on acid-free paper

This Birkhäuser imprint is published by Springer Nature
The registered company is Springer International Publishing AG Switzerland

Preface

This is a survey of the parabolic Anderson model (PAM), the Cauchy problem for the heat equation with random potential. This model and many variants and related models are studied for decades by many authors from various points of view and with various intentions. The PAM has rich and deep connections with questions on random motions in random potential, trapping of random paths, branching processes in random medium, spectra of random operators, Anderson localisation and more. We are mainly interested in the long-time behaviour of the solution of the PAM, which shows interesting behaviours like intermittency, mass concentration, ageing, Poisson process convergence of eigenvalues and eigenfunction localisation centres and more. Its mathematical investigations require combinations of tools from various parts of probability and analysis, like spectral theory of random operators, large deviations or extreme value statistics.

The research on the PAM and its variants has a high intensity since 1990 and continues to have. I felt that a survey text should be very useful, at a point at which most of the understanding of the basic model has been rigorously derived, and a variety of variants and additional features, like random environments and time-dependent potentials, and a number of related questions, like critically rescaled potentials, transition from concentrated to homogenised behaviour, spatial branching processes in random environment and Anderson localisation, receive an increasing interest.

The focus of this book is characterised by the intersection of a number of features, whose most important ones are the following:

- The solution to the PAM admits explicit formulas (Feynman-Kac formula and eigenvalue expansion),
- Its large-time behaviour can be investigated with the help of large-deviations theory,
- The arising variational formulas admit a deeper study and
- There are deep connections with the spectral theory for a prominent random Schrödinger operator, the Anderson operator.

All these aspects are more or less closely connected with the main property of the PAM, the intermittency, a concentration property of the main part of the solution in small islands. Intermittency is one of the leading ideas in this book and is almost ubiquitous.

For this reason, such important topics as directed polymers in random environment, PAM with drift and PAM with certain types of time-dependent potentials do not receive the space that they otherwise should have; they are just outside of the scope of this book. Actually, this text ends at a point where it is getting really interesting, as the stochastic heat equation and the KPZ equation come into play (however, the account on PAM with time-dependent potential given in Chap. 8 is quite comprehensive in a sense).

My intention was to provide a concise, but fairly complete, survey of the heuristic understanding of the PAM on one hand and of the state of the art of its mathematical treatment on the other hand. The goal is to quickly guide the reader to a good understanding of the essentials. I tried to give illuminating and nontechnical explanations, and I sometimes decided to provide simplified versions of the main theorems, many of which are embedded in the running text. Where some background is needed, the underlying theory is summarised in a most compact way, just at a length that is necessary to understand the fundamentals' all important connections.

There are a lot of precise references given to the first-hand literature, and many side remarks hint at deeper results and open problems that emanate from the material. I also found it useful to isolate the essentials of proof methods from the original papers, if time has shown that they are useful and can be adapted to several situations; not only Chap. 4 is devoted to this but also a number of remarks that are scattered over the text.

Originally, the text was meant to address experienced researchers, but in the course of writing, I felt that it would be desirable to attract also newcomers and young researchers to this field; therefore I added also explanations of terms, concepts, jargons and methods that are known to the community of the PAM and neighbouring fields. I hope that I found a style that is understandable and encouraging for all mathematically interested people from advanced undergraduates onwards.

In an appendix, I enumerated some open research directions that lie within the scope of this book or at its outer boundary. Certainly their choice relies on my personal taste, but I think that they each give rise to exciting new research, and hopefully they attract new people to the field.

Let me express my sincere thanks to my former PhD student Tilman Wolff, who helped me at an early stage in collecting some material, and to my current PhD student Franziska Flegel, who produced instrumental illustrations.

Berlin, Germany — Wolfgang König
March 2016

Some Remarks on Notation

For describing asymptotic assertions, I will use the symbol '$\sim$' to denote asymptotic equivalence, i.e. that the quotient of the two sides converges to one, '$\asymp$' for asymptotic comparability, i.e. that the ratio stays bounded and bounded away from zero, and $\ll$ and $\gg$ for asymptotic negligibility, i.e. that the ratio vanishes, respectively, and diverges to ∞. Furthermore, I use the Landau symbols $o(a_n)$ for quantities whose ratio with a_n vanishes asymptotically and $O(a_n)$ for positive quantities whose ratio with a_n stays bounded as $n \to \infty$. When I do not want to specify the sense of the asymptotic approximation, then I use the symbol '$\approx$', but often I indicate in words what I would like to mean by that. For expressing convergence, I often use the arrow $\to$, or $\stackrel{t\to\infty}{\longrightarrow}$, if I want to indicate the limiting parameter. Convergence of random variables in distribution or weak convergence of measures is written using $\Longrightarrow$.

For integrals and inner products both on $\mathbb{R}^d$ and on $\mathbb{Z}^d$, I use the brackets $\langle\cdot,\cdot\rangle$, e.g. $\langle\mu,f\rangle = \langle f,\mu\rangle = \int f \,\mathrm{d}\mu$ for functions f and measures μ or $\langle f,g\rangle = \int f(x)g(x)\,\mathrm{d}x$ for two functions f and g, which I sometimes also abbreviate as $\int fg$ if the domain is $\mathbb{R}^d$ and $\langle f,g\rangle = \sum_{z\in\mathbb{Z}^d} f(z)g(z)$ if it is $\mathbb{Z}^d$. For $p \in [1,\infty)$ and $B \subset \mathbb{Z}^d$, we denote by $\ell^p(B)$ the vector space of functions $f\colon B \to \mathbb{R}$ such that $\|f\|_p^p = \sum_{z\in B} |f(z)|^p$ is finite and $\|f\|_p$ is the norm of f.

For the parameter of some frequently used functions or processes, I use both the index notation and the bracket notation, depending on how much space the parameter requires. Hence a scale function $\alpha(t)$ may be also written α_t, and the simple random walk at time t is denoted both by X_t and by $X(t)$. Likewise, I write both $\mathbb{1}_A$ and $\mathbb{1}A$ for the indicator function on an event A.

Contents

Chapter 1
Background, Model and Questions

In this chapter, we introduce to the main subject of this book in Sect. 1.1. At its end, we give some guidelines to the remainder of this text.

1.1 Introduction and Scope of This Text

Random motions in random media or, closely related, partial differential equations with random coefficients, are an important subject on the edge between probability theory and differential equation theory, since there are a lot of applications to real-world problems in the sciences, like astrophysics, magnetohydrodynamics, chemical reactions, electrical networks, and much more. Diffusions of many kinds in disordered media play an important rôle in many physical applications, for example powders, porous earth and rocks, amorphous solids, atmospheric dust, micellar systems, and polymers in a solution. Physical examples include electron-hole recombination in random surfaces and in amorphous solids, exciton trapping and annihilation, and luminescence; see [HarBenAvr87] for more on applications and physical theories. For these reasons and also because of its mathematical interest, they have been studied a lot for decades, with a particular intensity in the last 20 years. There is a variety of models and hence a variety of questions, and of mathematical methods, concepts and theories.

In this text, we give a survey of the mathematical treatment of the *parabolic Anderson model (PAM)*, the (Cauchy problem for the) *heat equation with random potential*, a fundamental partial differential equation with random coefficients. A great advantage of this model is that the theory of its solutions is very well developed and is based on fundamental and important concepts like the Feynman-Kac formula and the spectral decomposition for compact operators; in particular it is rather explicit. Furthermore, the PAM exhibits a lot of cross-connections to other topics, like to the random walk in a random potential and the phenomenon of Anderson

W. König, *The Parabolic Anderson Model*, Pathways in Mathematics,
DOI 10.1007/978-3-319-33596-4_1

localisation. Furthermore, as outlined in [CarMol94], the PAM has a number of real-world applications and interpretations in various areas, like for spatial branching processes in a random environment, hydrodynamics and Burger's equation, and advection-convection equations for temperature fields.

The most prominent property of this model is a strong localisation effect, called *intermittency*: the random walk has a strong tendency to be confined to some few, small islands in the random medium, which are widely spread. These islands are very important, as the main mass of the solution is built up in them. Therefore, the global properties of the system are not determined by an averaging behaviour that comes from a law of large numbers or a central limit theorem. Hence techniques and assertions from (stochastic) homogenisation theory do not work. Instead, the behaviour of the system is determined by the local extremes of the random potential. Appropriate techniques come from the theories of large deviations, variational analysis, spectral theory, and extreme value statistics. This makes the PAM an exciting field to study from a mathematical perspective, as a lot of mathematical theories come together here, notably from probability and analysis.

For the study of the PAM, since 1990 a significant number of mathematical tools has been developed and adapted, which have later found useful also for the study of a number of other models in statistical physics. Furthermore, because of its relative simplicity and the extraordinary explicitness of representations of the solution, the PAM serves as one prime example of a partial differential equation with highly irregular behaviour, for which a detailed rigorous analysis of its solution is possible, going far beyond questions like existence and uniqueness. Intermittency was phenomenologically discussed for various other types of models [Zel84, ZelMolRuzSok87, ZelMolRuzSok88] because of its importance for applications in magnetohydrodynamics and, in particular, in the investigation of the induction equation with incompressible random velocity fields. However, the investigation of intermittency for the PAM began its mathematical career not before the seminal paper [GärMol90] and was very successful since then.

The PAM has become a popular model among probabilists and mathematical physicists, since there are a lot of interesting and fruitful connections to other interesting topics such as branching random walks with random branching rates, large-deviations analysis, spectra of random Schrödinger operators, extreme value statistics, convergence of point processes, variational problems. The mathematical activity on the PAM is on a high level, and many specific and deeper questions and variants were studied especially in the last few years, including, but not being limited to, time-dependent potentials, connections with Anderson localisation, transitions between quenched and annealed behaviour and PAM in a random environment. For this reason, it seems rather appropriate to provide a survey that collects and structures the relevant investigations and their interrelations, and to put them into a unifying perspective.

We decided to devote most attention to the case of a *static* random potential, i.e., a potential that does not depend on time, although there are many good reasons to study also the case of a time-dependent potential, a dynamic potential. For our decision, there are a couple of reasons, the most prominent of which are that (1) the static case has, in contrast to the dynamic case, many connections with the spectral properties of the Anderson Hamilton operator, and (2) the results that have been derived in the static case are much more explicit and more directly interpretable than in the dynamic case. Nevertheless, the set of time-dependent potentials that are interesting for the PAM is definitely much richer and comes from more different interesting applications and is still growing. We give an account on that in Chap. 8.

We also decided to put the main weight on the PAM in the *discrete* spatial setting, i.e., on $\mathbb{Z}^d$, even though the spatially continuous setting on $\mathbb{R}^d$ (i.e., with random walks replaced by Brownian motion) is equally interesting and mathematically challenging. The main reason for this is the existence of the formidable monograph on the $\mathbb{R}^d$-case, [Szn98], which is developed from the viewpoint of random path measures for Brownian motion trapped in a Poisson field of obstacles and contains a lot of results that have a direct impact to the PAM. In these notes, we will by no means neglect the material of [Szn98], but rephrase it in a way that is relevant for the PAM.

Let us give some guiding lines to the structure of this text. In Sect. 1.2 we introduce the model and the relevant questions, enumerate our main questions in Sect. 1.3, explain the fundamental phenomenon of intermittency heuristically in Sect. 1.4 and present the most important examples of random potentials in Sect. 1.5. In Chap. 2 we survey the most important tools, both probabilistic (e.g., the Feynman-Kac formula) and analytic (e.g., the eigenvalue expansion). One of the most fundamental questions, the asymptotics of the moments of the total mass of the model, is heuristically explained in Chap. 3: first we reveal the mechanism, then we bring and comment some detailed formulas. The most successful proof strategies are briefly explained in Sect. 4; they are useful also for the treatment of other models. The almost sure asymptotics of the total mass is explained in Chap. 5. Again, we first clarify the mechanism and provide explicit formulas afterwards. Chapter 6 is devoted to a detailed and deep investigation of the intermittency effect, that is, to very pronounced concentration properties of the model, and to resulting ageing properties. In Chap. 7, we summarize a number of additional directions that have been studied recently, like acceleration and deceleration, transition from quenched to annealed setting, and the PAM in a random environment. In the final Chap. 8, we enter the big and largely unexplored world of time-dependent potentials and describe some of the most relevant examples, motivations, results and open questions. The mathematical theory of the PAM has been rounding off since a few years, and for most of the phenomena adequate proof methods have been derived. However, a number of open questions in the scope covered by this text are still left, and we tried out best to concisely describe them in an encouraging manner in an appendix, in the hope that new researchers will be attracted to taking up some research in this field.

1.2 The Parabolic Anderson Model

Let us introduce the model in the spatially discrete case. We consider the non-negative solution $u: [0,\infty) \times \mathbb{Z}^d \to [0,\infty)$ to the *Cauchy problem* for the *heat equation with random coefficients* and localised initial datum,

The parabolic Anderson model (PAM)

$$\frac{\partial}{\partial t}u(t,z) = \Delta^{\rm d} u(t,z) + \xi(z)u(t,z), \qquad \text{for } (t,z) \in (0,\infty) \times \mathbb{Z}^d, \quad (1.1)$$

$$u(0,z) = \delta_0(z), \qquad \text{for } z \in \mathbb{Z}^d. \quad (1.2)$$

On the right-hand side of (1.1), we see the two main ingredients, a *random potential* $\xi = (\xi(z))_{z\in\mathbb{Z}^d}$ with values in $[-\infty,\infty)^{\mathbb{Z}^d}$, and the discrete *Laplace operator*,

$$\Delta^{\rm d} f(z) = \sum_{y\in\mathbb{Z}^d : y\sim z} \big[f(y) - f(z)\big], \qquad \text{for } z \in \mathbb{Z}^d, f: \mathbb{Z}^d \to \mathbb{R}. \quad (1.3)$$

Here $y \sim z$ means that y and z are nearest neighbours, i.e., the ℓ^1-norm of their difference is one. We will also sometimes write ∂_t instead of $\frac{\partial}{\partial t}$ for the partial derivative w.r.t. t. Certainly, $\Delta^{\rm d}$ applies only to the spatial dependence of u; we understand $\Delta^{\rm d} u(t,z)$ as $[\Delta^{\rm d} u(t,\cdot)](z)$. Sometimes the Laplace operator is defined without the term $-f(z)$, which differs from $\Delta^{\rm d}$ just by adding $2d$ times the identity operator. Many other authors put $1/2d$ or a new parameter (often 'κ') as a prefactor. The operator $\Delta^{\rm d}$ in (1.3) is the generator of a simple random walk on $\mathbb{Z}^d$ in continuous time that makes independent and identically distributed (henceforth abbreviated by 'i.i.d.') steps after independent exponential times with parameter $2d$ and expectation $1/2d$. Even though the random walk is, strictly speaking, absent in (1.1), it will appear in many explanations, arguments and proofs. It is called the *random walk in random potential*. Another name for (1.1) is the *heat equation with random potential*, as one of its interpretations is the description of a heat flow through a random medium that is called a potential.

For most of this text, the random potential ξ is assumed to be i.i.d. without further mentioning, that is, the collection of variables $\xi(z)$ with $z \in \mathbb{Z}^d$ is independent, and they are identically distributed. (However, see Sect. 7.2 for correlated potentials.) We denote expectation and probability with respect to ξ by $\langle\cdot\rangle$ and Prob, respectively. The notation $\langle\cdot\rangle$ for averages over media is common in mathematical physics, and one should indeed consider ξ as a random medium.

Remark 1.1 (The Anderson operator) The operator $\Delta^{\rm d} + \xi$ appearing on the right-hand side of (1.1) is called the *Anderson Hamiltonian* or *Anderson operator*; its spectral properties close to the top of its spectrum will be very important for us in the sequel, see Remark 2.2.3. This operator is one of the most prominent examples of a *random Schrödinger operator.*

If the Anderson operator on the right-hand side of (1.1) is multiplied with the imaginary unit, then an equation arises that is often called the (time-dependent) *Anderson Schrödinger equation* or the (time-dependent) *Schrödinger equation with random potential*:

$$\partial_t u = \mathrm{i}\Delta^{\mathrm{d}} u + \mathrm{i}\xi u \qquad \text{on } [0,\infty) \times \mathbb{Z}^d. \tag{1.4}$$

It is of basic importance in non-relativistic quantum mechanics and motivates the high interest in the spectral properties of $\Delta^{\mathrm{d}} + \xi$ in any part of its spectrum (not only close to the top). It describes the time evolution of the complex-valued wave-function, which represents the quantum state of the electron in the potential ξ. Equation (1.4) was established by Erwin Schrödinger [Sch26]. The name 'parabolic Anderson problem' for (1.1) points out the fundamental difference between the two equations in (1.1) and (1.4). Equation (1.1) is much more amenable to a probabilistic analysis than (1.4), and the spectrum enters its large-time analysis only close to its top. However, recently a possibility was found to interpret the solution to (1.4) in terms of branching processes and to make it amenable to a probabilistic analysis, see Remark 2.2. ◇

Here is the fundamental starting point for the study of the PAM.

Theorem 1.2 (Existence and uniqueness) *Almost surely, the equation* (1.1)–(1.2) *has precisely one non-negative solution* $u(t,\cdot)$ *if the potential satisfies the integrability condition*

$$\Big\langle \Big(\frac{\xi(0) \vee 2}{\log(\xi(0) \vee 2)} \Big)^d \Big\rangle < \infty, \tag{1.5}$$

where $x \vee y$ *is the maximum of* x *and* y.

See [GärMol90, Theorem 2.1] and [CarMol94] for a proof of this fact and the derivation of the *Feynman-Kac formula* for the solution, which we spell out in Sect. 2.1.2. Note that in [GärMol90] there is a flawed formulation of (1.5) with $\xi(0)_+/(\log(\xi(0)))_+$ instead of $(\xi(0) \vee 2)/\log(\xi(0) \vee 2)$, where x_+ denotes the positive part of x. This creates unwanted trouble with values of $\xi(0)$ in $(0,1]$. The only meaning of (1.5) is to upper-bound the extremely large values of $\xi(0)$.

It is also shown in [GärMol90] and [CarMol94] that the condition (1.5) is necessary in a certain sense. The main argument for the existence part is that the Feynman-Kac formula is shown to be finite (using a comparison of the speed of the underlying random walk and the asymptotic growth of the potential), and this implies that this formula is a solution to (1.1)–(1.2). The uniqueness part is done by showing that, for some sufficiently negative α, the level set $\{z \in \mathbb{Z}^d : \xi(z) \leq \alpha\}$ does not contain any unbounded connected component.

Henceforth, we assume that (1.5) is satisfied and denote by u the non-negative solution. See Remark 1.5 for other initial conditions instead of (1.2), and see Remark 1.6 for the PAM in finite subsets of $\mathbb{Z}^d$ with boundary conditions. We refer the reader to [Mol94], [CarMol94] and [GärMol90] for more background and more details of basic mathematical properties of the model, and to [GärKön05] for a survey on some mathematical results up till 2005.

Remark 1.3 (Interpretation) The PAM describes a random particle flow in $\mathbb{Z}^d$ through a random field of sinks and sources. Sites z with $\xi(z) < 0$ are interpreted as sinks, traps or obstacles ('hard' for $\xi(z) = -\infty$ and 'soft' for $\xi(z) \in (-\infty, 0)$), while sites z with $\xi(z) \in (0, \infty)$ are called sources and those with $\xi(z) = 0$ neutral. There is an additional interpretation in terms of a branching process in a field of random branching rates, see Sect. 2.1.1. ◇

Remark 1.4 (Between smoothness and roughness) Two competing effects are present: the diffusion mechanism governed by the Laplacian, and the local growth governed by the potential. The diffusion tends to make the random field $u(t, \cdot)$ flat, whereas the random potential ξ has a strong tendency to make it irregular. This is understood best by considering the two separate equations

$$\partial_t u(t, z) = \Delta^{\mathrm{d}} u(t, z) \qquad \text{and} \qquad \partial_t u(t, z) = \xi(z) u(t, z)$$

under the same initial condition. The first one is called the (Cauchy problem for the) *heat equation* and implies that the exponential growth rate of $u(t, z)$ at some point $z \in \mathbb{Z}^d$ is proportional to the sum $\sum_{y \sim z}[u(t, y) - u(t, z)]$. In particular, $u(t, z)$ grows if the average value of u in the neighbouring points is higher than $u(t, z)$ itself and decreases in the opposite case, which corresponds to heat spreading evenly over a surface. The second equation admits the simple solution $u(t, z) = \mathrm{e}^{t\xi(z)}$, $z \in \mathbb{Z}^d$, which does not admit any interaction between different lattice points and is extremely irregular for large t as we may have considerably different growth rates in different lattice points. In (1.1), both these effects interact; the Laplacian smears the extreme roughness coming from the irregularity of the potential. ◇

Remark 1.5 (Other initial conditions) Instead of the localised initial condition $u(0, \cdot) = \delta_0(\cdot)$ in (1.2), certainly also other initial conditions $u(0, \cdot) = u_0(\cdot)$ may be considered, as long as the initial function u_0 is non-negative and satisfies

$$\limsup_{z \to \infty} \frac{(\log u_0(z))_+}{|z| \log |z|} < 1; \tag{1.6}$$

see [GärMol90, Theorem 2.1]. Observe that the symmetry of the Laplace operator implies that the *superposition principle* holds: If $u(t, \cdot)$ and $\tilde{u}(t, \cdot)$ are the solutions with initial condition u_0 and $\tilde{u}_0$, respectively, then $(u + \tilde{u})(t, \cdot)$ is the solution with initial condition $u_0 + \tilde{u}_0$. The most-studied choice, apart from $u_0 = \delta_0$, is the homogeneous initial condition $u_0 \equiv 1$, in which case we write v for the solution to (1.1); then the random field $v(t,\cdot)$ is *stationary*, i.e., its distribution is shift-invariant for any t. We will always denote the solution with localised initial condition δ_0 by u, as in (1.1)–(1.2). Then the superposition principle implies that $v(t,0) = \sum_{z\in\mathbb{Z}^d} u(t,z)$. ◇

Remark 1.6 (The PAM in boxes) The PAM can also be considered in a given finite set $B \subset \mathbb{Z}^d$, but one has to specify the boundary condition. The two mainly used boundary conditions are the *Dirichlet boundary condition* (by which we mean *zero boundary condition*) and the *periodic boundary condition*, the latter only for the case that B is a rectangle. We will then take B always as a cube, often a centred cube.

If B contains the origin, we denote by $u_B\colon (0,\infty) \times \mathbb{Z}^d \to [0,\infty)$ the localised solution with zero boundary condition, i.e., the solution to (1.1)–(1.2) on $(0,\infty)\times\mathbb{Z}^d$ under the extra requirement that $u(t,z) = 0$ for every $z \in \mathbb{Z}^d \setminus B$. Note that in the term $\Delta^{\mathrm{d}} u(t,\cdot)$ also bonds between B and B^{c} occur.

If $B = (-R,R]^d \cap \mathbb{Z}^d$ with $R \in \mathbb{N}$ is a centred cube, then we denote by $u_B^{(\mathrm{per})}\colon (0,\infty) \times B \to [0,\infty)$ the solution to (1.1)–(1.2) with periodic boundary condition. There are two ways to understand this definition. First, one conceives $u_B^{(\mathrm{per})}$ as the solution to (1.1)–(1.2) on $(0,\infty) \times \mathbb{Z}^d$ with the extra condition of periodicity, i.e., $u(t,z+R\mathrm{e}_i) = u(z)$ for any $t \in (0,\infty)$, $z \in \mathbb{Z}^d$ and $i \in \{1,\dots,d\}$, where $\mathrm{e}_i \in \mathbb{Z}^d$ is the i-th unit vector, and restricts this solution to $(0,\infty) \times B$. Second, one restricts (1.1) to $z \in B$ and replaces the Laplace operator Δ^{d} by the one on $\ell^2(B)$ with periodic boundary condition. This operator is actually nothing but the canonical Laplace operator on the d-dimensional torus B, which is defined by defining any site with at least one of the d components equal to R as a neighbour of the site with this component replaced by $-R+1$.

Both u_B and $u_B^{(\mathrm{per})}$ are important for the study of the PAM, as they will turn out to serve as lower, respectively upper, bounds for u, see Sect. 2.1.3. ◇

Remark 1.7 (The PAM on $\mathbb{R}^d$) The spatially continuous version of the parabolic Anderson model is given by

$$\frac{\partial}{\partial t}u(t,x) = \Delta u(t,x) + V(x)u(t,x), \qquad \text{for } (t,x) \in (0,\infty) \times \mathbb{R}^d, \tag{1.7}$$

$$u(0,x) = \delta_0(x), \qquad \text{for } x \in \mathbb{R}^d, \tag{1.8}$$

where $\Delta = \sum_{i=1}^d (\frac{\partial}{\partial x_i})^2$ is the usual Laplace operator, and $V\colon \mathbb{R}^d \to [-\infty,\infty)$ is a random field, which we assume to be sufficiently regular and integrable. If V is stationary (i.e., if its distribution is invariant under shift by any vector in $\mathbb{R}^d$) and the localised initial condition $u(0,\cdot) = \delta_0$ is replaced by the condition $u(0,\cdot) = 1$, then also $u(t,\cdot)$ is a stationary field for any t.

All the preceding has an analogue; we are not going to spell out all this explicitly. We refer to [Szn98] for the fundamental theory concerning (1.7). See Sect. 2.1.2 for a formulation of the Feynman-Kac formula. ◇

Remark 1.8 (The PAM on graphs) It makes perfect sense to consider the PAM on an arbitrary graph G instead of $\mathbb{Z}^d$, replacing Δ^d by the standard graph Laplacian

$$\Delta\varphi(g) = \sum_{h\in G:(g,h) \text{ is an edge}} (\varphi(h) - \varphi(g)).$$

One interesting choice is the graph $G = \{0,1\}^N$ for some $N \in \mathbb{N}$, which models the set of all gene sequences of length N (where we simplify the presence of four alleles to just two). For this choice, the branching process picture that we explain in Sect. 2.1.1 makes good sense for biological applications, as it models the random occurrences of mutants in a large population. The state space G is not interpreted as the region in which the population lives, but the gene pool that the individuals may have. The potential $\xi: G \to \mathbb{R}$ is the 'fitness landscape', which attaches to each sequence g its fitness $\xi(g)$. Choosing the $\xi(g)$s as independent and identically distributed random variables is currently one of the most popular choices in the modeling of biological systems.

An interesting open question is how much time (in dependence of the length of the gene sequences, N) the system needs to reach the 'fittest' site with the main bulk of the population. This question was answered for the complete graph $\{1, 2, \dots, N\}$, where all bonds $\{i, j\}$ are edges with $i \neq j$, in [FleMol90]. ◇

1.3 Main Questions

Let us formulate our main questions about the PAM. Generally, we would like to describe the solution $u(t,\cdot)$ asymptotically as $t \to \infty$. One of the main objects of interest is the *total mass* of the solution,

$$U(t) = \sum_{z\in\mathbb{Z}^d} u(t,z), \qquad \text{for } t > 0. \tag{1.9}$$

We ask the following questions:

- **Asymptotics.** What is the asymptotic behaviour of $U(t)$ as $t \to \infty$?
- **Localisation.** What are the regions that contribute most to $U(t)$? How do these regions depend on the potential and how on time? How many of them are there and where do they lie?
- **Shapes.** What do the typical shapes of the potential $\xi(\cdot)$ and of the solution $u(t,\cdot)$ look like in these regions?
- **Mass flow and ageing.** What is the behaviour of the entire process of the mass flow, $(u(t,\cdot))_{t\in[0,\infty)}$? Does it exhibit ageing properties?

As the above questions suggest, we do not expect a homogeneous behaviour of the solution, but a localised one, i.e., a concentration of $u(t,\cdot)$ on relatively few islands, with a peculiar form of the potential and the solution inside these islands. This will be highlighted in Sect. 1.4. While the first three questions concern a snapshot of the solution at late times, the last one refers to the flow of the mass over time intervals, the entire evolution of the process. Here we would like to understand the correlations of u at several different times or even over an entire time interval.

Chapters 3, 5 and 6 will give the answers to the above questions; some preliminary, heuristic answers will be discussed in Sect. 2.3.

As is common in statistical mechanics, we distinguish between the so-called *quenched* setting, where we consider $u(t,\cdot)$ almost surely with respect to the medium ξ, and the *annealed* one, where we average with respect to ξ. These notions stem from metallurgy and are frequently used in probability theory and statistical physics, even though they are generally not defined in a universal way; in some situations their precise meaning is due to the author's point of view and intention.

There are more settings in which asymptotic results can be derived: *convergence in probability* and *convergence in distribution*. They are sometimes applied in cases in which the annealed setting does not exist because of the non-existence of the expected value; we will rely on them in cases in which we will not be able to take expectations w.r.t. ξ but the derivation of almost-sure assertions is technically cumbersome.

It is clear that the quantitative properties of the solution strongly depend on the distribution of the field ξ (more precisely, on the upper tail of the distribution of the random variable $\xi(0)$), and that different phenomena occur in the different settings.

1.4 Intermittency

The long-time behaviour of the parabolic Anderson problem is well-studied in the mathematics and mathematical physics literature because it is an important example of a model exhibiting an *intermittency effect*. Here is a heuristic definition:

> **Intermittency:**
> The random time-dependent function u is called *intermittent* if $u(t,\cdot)$ develops, for large t, high peaks on few, small, remote islands, the **intermittent islands**, which carry most of the total mass $U(t) = \sum_{z\in\mathbb{Z}} u(t,z)$.

This captures the geometric essence of this effect; one can call it also *geometric intermittency*. Many of the investigations of the PAM were motivated by a desire to understand it in detail, and it will serve us as a leading idea. Much of the following, in particular Chaps. 3 and 6, is devoted to a thorough explanation of intermittency in this spatial sense.

However, this definition of intermittency is on one hand too detailed to be formulated concisely and on the other hand too little rigorous, as a precise definition would have to depend on details of the potential. Hence, less detailed, but rigorous definitions of intermittency are helpful. One of the most often used definitions is in terms of the moments of $U(t)$:

Moment intermittency:
The random time-dependent function u is called *intermittent* if its total mass $U(t)$ satisfies

$$\limsup_{t\to\infty} \frac{1}{t} \log \frac{\langle U(t)^p\rangle^{1/p}}{\langle U(t)^q\rangle^{1/q}} < 0, \qquad \text{for } 0 < p < q. \tag{1.10}$$

Here we recall that $\langle \cdot \rangle$ denotes expectation with respect to ξ. The left-hand side is non-positive by Jensen's inequality; the requirement is that the quotient of the p-norm and the q-norm decays even exponentially fast in t.

Let us briefly illustrate what (1.10) implies for the large-time behaviour of the solution, see [GärMol90, Section 1]. Write $-a$ for the left hand side of (1.10), pick some $\varepsilon \in (0, a)$, and consider the extreme event

$$E_t = \{U(t) > \mathrm{e}^{t\varepsilon}\langle U(t)^p\rangle^{1/p}\}$$

that the total mass is exponentially much larger than its p-norm. With Prob denoting the probability w.r.t. the random potential ξ, we may estimate

$$\mathrm{Prob}(E_t) = \mathrm{Prob}\Big(\frac{U(t)^p}{\langle U(t)^p\rangle} > \mathrm{e}^{tp\varepsilon}\Big) \le \mathrm{e}^{-tp\varepsilon}, \tag{1.11}$$

with the help of Markov's inequality. Hence, E_t is an exponentially rare event. On the other hand, we see that the main contribution of the q-th moment comes from this rare event as follows. Indeed, we have

$$\frac{\langle U(t)^q \mathbb{1}_{E_t^c}\rangle}{\langle U(t)^q\rangle} = \frac{\langle U(t)^q \mathbb{1}\{U(t)^q \le \mathrm{e}^{tq\varepsilon}\langle U(t)^p\rangle^{q/p}\}\rangle}{\langle U(t)^q\rangle} \le \frac{\mathrm{e}^{tq\varepsilon}\langle U(t)^p\rangle^{q/p}}{\langle U(t)^q\rangle}. \tag{1.12}$$

(We write $\mathbb{1}_A$ and $\mathbb{1}A$ for the indicator function on an event A, if no confusion can arise.) Combining with (1.10), we see that its exponential rate is negative:

$$\limsup_{t\to\infty} \frac{1}{t} \log \frac{\langle U(t)^q \mathbb{1}_{E_t^c}\rangle}{\langle U(t)^q\rangle} \le q\varepsilon - qa < 0.$$

This means that the left-hand side of (1.12) decays exponentially fast towards 0, which implies that $\langle U(t)^q\rangle \sim \langle U(t)^q \mathbb{1}_{E_t}\rangle$. Summarizing, moment intermittency implies that the main contribution of the qth moment comes from an event whose probability decays exponentially fast and does not contribute to the pth moment, if $0 < p < q$.

Strictly speaking, (1.10) does not say anything about the spatial structure of the solution $u(t,\cdot)$. However, if one recalls from Remark 1.5 that $U(t) = v(t,0)$ with v the solution of (1.1) with homogeneous initial condition $v(0,\cdot) \equiv 1$, then we see from (1.11), using the ergodic theorem, that the set $\{z \in \mathbb{Z}^d : v(t,z) > e^{t\varepsilon}\langle U(t)^p\rangle^{1/p}\}$ of highest exceedances of the field $v(t,\cdot)$ has an exponentially small density. What is not clear at the moment (and whose formulation needs also some more care) is that it is this set that gives the main contribution to the total mass $U(t)$, more precisely, to its q-th moments; actually it is a set that contains the intermittent islands.

Intermittency is an effect that is indeed present in the PAM in great generality. One of the starting points of the interest in the PAM and of the research on the PAM is the following fact, see [GärMol90, Theorem 3.2].

Theorem 1.9 *If* ξ *is truly random, the PAM is intermittent in the sense that* (1.10) *holds.*

This fact is one of the leading sources of motivation and has been severely extended into various directions; much of this text is devoted to this.

1.5 Examples of Potentials

Let us present and comment on the most interesting examples of random potentials that have been studied in the literature in connection with the PAM.

We will be almost entirely be interested in phenomena that arise in the limit of late times, $t \to \infty$. Let us stress that these phenomena crucially and almost entirely depend on the behaviour of the upper tails, i.e., the asymptotics of the function $r \mapsto \mathrm{Prob}(\xi(0)) > r)$ as $r \to \infty$. This will be substantiated in Sect. 2.3.1. Therefore, it is often just this asymptotics that is specified, not the entire distribution of the potential. A first main distinction is whether or not the essential supremum

$$\mathrm{esssup}\,(\xi(0)) = \sup\{r \in \mathbb{R} : \mathrm{Prob}(\xi(0) > r) < 1\}$$

is equal to ∞ or not. In the latter case, it is no restriction of the generality to assume that $\mathrm{esssup}\,(\xi(0)) = 0$, as the transition from ξ to $\xi + c$ with a constant c means a transition from $u(t,\cdot)$ to $u(t,\cdot)e^{ct}$ for the solution. We call the distribution of $\xi(0)$ the *single-site distribution*; however, note the terminology used in Example 1.19. Each single-site distribution comes with its *logarithmic moment generating function*, sometimes also called the *cumulant generating function*,

$$H(t) = \log\langle e^{t\xi(0)}\rangle, \tag{1.13}$$

whose large-t asymptotics stand in a one-to-one connection with the upper tails of $\xi(0)$. We will pay special attention to this function, as it is used often in proofs.

Recall from Remark 1.3 the classification of sites z as a hard trap if $\xi(z) = -\infty$, a soft trap if $\xi(z) \in (-\infty, 0)$, neutral if $\xi(z) = 0$ and a source if $\xi(z) \in (0, \infty)$.

1.5.1 Discrete Space

By far most of the examples of random potentials $\xi = (\xi(z))_{z\in\mathbb{Z}^d}$ that we treat in this text consist of independent and identically distributed (i.i.d.) random variables. In principle, any single-site distribution (i.e., marginal distribution in one site) that satisfies the condition (1.5) is interesting for a consideration of the PAM, but some received more or less interest for various reasons. Here we list some important single-site distributions and give some few remarks about the main properties of the PAM with the respective distribution. In particular, we establish borders between certain potential classes, which will later turn out to exhibit characteristic properties.

The case of correlated potentials is nevertheless also highly interesting and has not been fully explored yet, but we decided to source this subject out to Sect. 7.2.

Example 1.10 (Bernoulli traps) The case when the field ξ assumes the values $-\infty$ and 0 only has a nice interpretation in terms of a *survival probability* (see Example 2.9) and is therefore of particular importance. The logarithmic moment generating function is $H(t) = \log\langle e^{t\xi(0)}\rangle = \log p$ for $t > 0$, where p is the probability that a given site is neutral, and $H(0) = 0$. See Example 1.11 for an embedding of this potential in a much larger class. For more about the PAM with Bernoulli trap potential, see Example 2.9, Remark 3.19, and Sect. 7.7. ◇

Example 1.11 (Other bounded potentials) As we will later see, it is of interest to extend the scope of bounded potentials, by which we actually mean potentials that are bounded *from above*. For such potentials, we assume without loss of generality that the essential supremum esssup $(\xi(0))$ is equal to zero. The relevant choice of parameters for determining the upper tails is the following. For some $D \in (0, \infty)$ and $\gamma \in (0, 1)$,

$$\text{Prob}(\xi(0) > -x) \approx \exp\{-Dx^{-\frac{\gamma}{1-\gamma}}\}, \qquad x \downarrow 0. \tag{1.14}$$

That is, we postulate a stretched-exponential behaviour with any negative exponent. The $\approx$-sign is to be interpreted as logarithmic equivalence, i.e., asymptotic equivalence of the logarithms. The logarithmic moment generating function behaves like $H(t) \sim -Ct^\gamma$ for some $C = C(D, \gamma)$. The somewhat strange way in which we incorporated γ in the power of x is motivated by an embedding in a larger class of potential distributions that we will discuss in Sect. 3.4 below; actually, this class is identical to the class (B) of that classification, see Remark 3.19. The boundary case $\gamma = 0$ contains the Bernoulli trap case of Example 1.10, but also more, for example potentials that attain the values -1 and 0 only and the uniform distribution on $[-1, 0]$. The latter distribution (more precisely, the uniform distribution on $[0, 1]$) is one of the main motivating examples for the study of the spectrum of the Anderson Hamiltonian $\Delta^{\mathrm{d}} + \xi$ in the community of Anderson localisation, as it has an interpretation in terms of an alloy of metals; $\xi(z)$ is here the percentage of the amount of one of the metals in z.

The boundary case $\gamma = 1$ is phenomenologically contained in the almost bounded potentials of Example 1.13.

Let us already reveal here that one characteristics of bounded potentials is that the intermittent islands are relatively large as functions of t; actually their radii turn out to diverge as a power of t in the annealed setting. ◇

Example 1.12 (The double-exponential distribution) Of high interest is also the single-site distribution given by

$$\mathrm{Prob}(\xi(0) > r) = \exp\big\{ - \mathrm{e}^{r/\rho}\big\}, \qquad r \in \mathbb{R}, \tag{1.15}$$

with parameter $\rho \in (0, \infty)$. The name *double-exponential distribution* refers to the right-hand side in a naive way; $\xi(0)$ has just a reflected Gumbel distribution. The logarithmic moment generating function is $H(t) = \log\langle \mathrm{e}^{t\xi(0)}\rangle = \rho t \log t + \rho t + o(t)$ for large t (see [GärMol98], e.g.). The importance of this distribution for the PAM comes from the fact that the intermittent islands (see Sect. 1.4) turn out to be discrete, i.e., their diameter not depending on t, in particular not growing with time, but still showing an interesting spatial shape. This makes it a distribution that is nice to study, since a great source of technical difficulties is absent. This potential is unbounded to infinity and produces high peaks in the solution $u(t, \cdot)$. In Sect. 3.4 it will turn out to be the main representative of the class (DE), see Remark 3.17.

The parameter ρ describes the thickness of the tails, i.e., the tendency of the potential to assume very high values: the larger ρ is, the easier it is for the potential to assume large values. This is reflected in the fact that the size of the intermittent islands is decreasing with ρ, as we will later see. We will also see later that the two boundary cases $\rho = 0$ and $\rho = \infty$ correspond to the almost bounded case of Example 1.13 and to the heavy-tailed case of Example 1.14, respectively.

Because of its central position with respect to the asymptotic picture of the PAM, the double-exponential distribution will be discussed with respect to many types of results, see Remarks 3.17 and 5.9 and Sects. 6.2, 6.3.2, 6.4.1 and 7.1.1. ◇

Example 1.13 (Almost bounded potentials) This is a class of single-site potentials that can be seen as interpolating between the bounded distributions of Example 1.11 for $\gamma = 1$ and the double-exponential distribution of Example 1.12 with $\rho = 0$. Indeed, one obtains examples of potentials (unbounded from above) by replacing ρ in (1.15) by a sufficiently regular function $\rho(r)$ that tends to 0 as $r \to \infty$, and other examples (bounded from above) by replacing γ in (1.14) by a sufficiently regular function $\gamma(x)$ tending to 1 as $x \downarrow 0$.

These potentials constitute another important class of potentials, the class (AB) of *almost bounded potentials*, see Remark 3.18. It turns out in [HofKönMör06] (see

Sect. 3.4) that the radius of the intermittent islands of the solution $u(t,\cdot)$ for the PAM with this potential diverges with $t \to \infty$ on a scale that interpolates between the bounded and the double-exponential case, as one may expect. Despite a somewhat tenacious introduction of examples of this class of distributions, this class has the very nice property that the potential and the solution in the intermittent islands take the shape of perfect parabolas and Gaussian densities, respectively. For further properties of the PAM with almost bounded potentials, see Remark 3.8. ◇

Example 1.14 (Heavier-tailed potentials) By this we mean potentials that are unbounded to ∞ and have heavier tails than the double-exponential distribution, i.e., phenomenologically the case $\rho = \infty$ from Example 1.12. Hence, it includes much more distributions than is usually summarized under the – somewhat unsharp – term 'heavy-tailed' distribution, which practically always implies the infiniteness of positive exponential moments. However, the class of heavier potentials comprises also the Weibull distribution $\mathrm{Prob}(\xi(0) > r) = \mathrm{e}^{-Cr^\alpha}$ with $\alpha > 0$ and the Gaussian distribution; note that all the positive exponential moments of the Weibull distribution with $\alpha > 1$ are finite. But also the Pareto distribution $\mathrm{Prob}(\xi(0) > r) = r^{-\beta}$ for $r \in [1,\infty)$ with $\beta > 0$ belongs to the heavier-tailed distributions; note that one has to assume that $\beta > d$, in order that the condition (1.5) is satisfied.

The heavier-tailed potentials constitute the class (SP) of potentials, see Remark 3.16. The 'SP' indicates that the intermittent islands turn out to be singletons in [GärMol98], i.e., the potential develops 'single peaks', see Sect. 3.4. Generally, the more heavily tailed the potential, the stronger pronounced the concentration effect is and also, on the technical side, the more easily proved.

The Weibull distribution with $\alpha < 1$ and the Pareto distribution are even so heavy-tailed that they do not have finite exponential moments, i.e., the function in (1.13) is not finite for $t > 0$. Accordingly, all the moments of the solution $u(t,z)$ are infinite, and the annealed setting does not exist. However, starting with [HofMörSid08, KönLacMörSid09, MörOrtSid11], distributional properties of $u(t,\cdot)$ and limit theorems in probability were derived, and the most detailed pictures that can currently be proved for the PAM were first derived for Pareto-distributed potentials; see for example the mass concentration property in Sect. 6.4.2 and ageing properties in Sect. 6.5. ◇

1.5.2 *Continuous Space*

Let us list and comment on examples of random potentials V on $\mathbb{R}^d$ for which the PAM in (1.7) is studied in the literature. Most of them have a quite high degree of regularity and have positive and sometimes infinite correlation length. The study of the PAM for entirely uncorrelated potentials is currently in its infancy; see Example 1.21. Almost all our examples are stationary, i.e., distributional invariant under spatial shifts by any vector in $\mathbb{R}^d$. One exception is the field that is equal to $\xi(z)$ on the box $z + [0,1)^d$ for any $z \in \mathbb{Z}^d$ with an i.i.d. potential $(\xi(z))_{z\in\mathbb{Z}^d}$, which is more or less an discrete-space potential. Such a field is a – somewhat artificial – example of a potential with finite, positive length of correlation; actually, one might see all the examples of i.i.d. potentials $(\xi(z))_{z\in\mathbb{Z}^d}$ of Sect. 1.5.1 in this light.

Example 1.15 (Poisson traps) One of the most-studied potentials is given in the form

$$V(x) = -\sum_i W(x-x_i) = -\int_{\mathbb{R}^d} W(x-y)\,\omega(\mathrm{d}y), \qquad x \in \mathbb{R}^d, \tag{1.16}$$

where $(x_i)_i$ or $\omega = \sum_i \delta_{x_i}$ is a Poisson point process in $\mathbb{R}^d$ with constant intensity $\nu \in (0,\infty)$, and $W\colon \mathbb{R}^d \to [0,\infty]$ is a fixed given non-negative function, called *cloud*. Since W is non-negative, we call V a *Poisson obstacles potential* or a *Poisson field of traps*, in contrast with the case of a non-positive cloud in Example 1.16. Canonical choices of non-negative clouds are $W = C\mathbb{1}_K$ for some compact set $K \subset \mathbb{R}^d$ (say, a centred ball) containing the origin and for some $C \in (0,\infty]$, or W some non-negative continuous function with compact support, or $W(x) = C|x|^{-q}$ for some $C \in (0,\infty)$ and $q \in (0,\infty)$. The random potentials obtained for these choices can be seen as the natural analogues to the non-positive potentials of Examples 1.10 and 1.11.

However, one must be careful, as, for $W(x) = C|x|^{-q}$ with $q \leq d$, the potential V is infinite almost everywhere (i.e., $V \equiv -\infty$), almost surely [CheKul12, Proposition 2.1]. For $q \in (d/2, d)$, one can still make a good sense of the PAM by considering a renormalised version, see Remark 2.6 for more on the potential and Sect. 7.3.4 for the large-t behaviour of the moments of the solution to the PAM.

See Example 2.10 for more on the PAM with Poisson obstacles potential. This model is also called the *Brownian motion among Poisson obstacles*. The large-t asymptotics of the moments of the solution to the PAM with cloud $W(x) = C|x|^{-q}$ and $q \in (d+2,\infty)$ are very similar to the ones with $W = C\mathbb{1}_K$; see Sect. 3.5.1. However, for $q \in (d, d+2)$, new phenomena arise; see Example 7.6. For the almost sure asymptotics of the PAM, see Sect. 5.2, and see Sect. 7.3.3 for compactly supported and bounded clouds with some t-dependent prefactor, and Sect. 7.10 for results on the PAM with cloud $W(x) = C\mathbb{1}_K(x)$ and some t-dependent prefactors in front of the potential. ◇

Example 1.16 (Poisson shot-noise potential) It makes perfect sense to choose the Poisson cloud in (1.16) with the other sign, in which case we write $\varphi\colon \mathbb{R}^d \to [0,\infty)$ instead of $-W$ and

$$V(x) = \sum_i \varphi(x-x_i), \tag{1.17}$$

where $(x_i)_i$ is a Poisson point process in $\mathbb{R}^d$ with constant intensity $\nu \in (0,\infty)$. Such a potential is sometimes called a *Poisson shot-noise potential*. (This name comes from the special choice of φ as a Gaussian density, say, where each part $\varphi(x_i+\cdot)$ represents the distribution of all the bullets of a shot noise that is fired at x_i.) The potential ξ, and hence solution $u(t, \cdot)$, can easily achieve very high values in areas where many Poisson points stand close together; then many copies of the cloud φ are superposed. See Sect. 3.5.3 for the annealed large-t behaviour of the solution to the PAM with Poisson shot-noise potential for bounded, smooth clouds φ, and Example 5.15 for the quenched behaviour.

If φ is assumed to be bounded and compactly supported, no problem arises in the definition of the potential and construction of the solution of the PAM, but the (very natural) choice $\varphi(x) = \theta|x|^{-p}$ with $p \in [0, d)$ causes problems, as $V(x) = \infty$ for almost all x, almost surely [CheKul12, Proposition 2.1]. See Remark 2.6 for a renormalised version. ◇

Example 1.17 (Gibbsian point potentials) In both Examples 1.15 and 1.16, instead of a Poisson random point field $(x_i)_i$, one can also pick the random field $(x_i)_i$ to be a Gibbsian point field, i.e., a point field that, unlike a Poisson process, has some nontrivial correlation between the particles. In [Szn93, Mer03] (see also [Szn98]), a Gibbsian point field is considered that arises from a non-homogeneous Poisson process in $\mathbb{R}^d$ via a symmetric pair-interaction potential by means of the DLR equation as the corresponding Gibbs measure in $\mathbb{R}^d$. The pair-interaction potential is assumed there as bounded from below, compactly supported and superstable. The details of the definition of such a process are a bit cumbersome, but rely on standard Gibbs measure theory. The choice of this potential is motivated by the wish to model the random matter in a more realistic way. See Remarks 3.22 and 5.12 for the results of [Szn93]. Under the above assumptions on the pair-interaction potential they actually do not differ much from the ones in the Poisson case, but the proofs are significantly more involved. However, in [Mer03] (see Remark 7.8), the Gibbsian potential is scaled in a critical manner, and some physical properties of the Gibbsian process enter the description of the long-time asymptotics of the PAM. ◇

Example 1.18 (Gaussian potentials) Another interesting and natural choice is to take V as a Gaussian field with sufficiently good regularity properties. A canonical assumption is twice continuous differentiability of the covariance function [GärKön00, GärKönMol00], in which case the potential has a modification that is Hölder continuous with any parameter in $(0, 1)$. Recently there was also some efforts to study the PAM under much less regularity assumptions [Che14], where the potential V is not even a function, but only a measure, and see Example 1.21 for the uncorrelated case. Furthermore, see Sect. 3.5.2 and Examples 5.13 and 7.5 for more specific questions about the PAM on $\mathbb{R}^d$ with Gaussian potential. ◇

Example 1.19 (Alloy-type potentials) One of the most-studied random potentials in the community of Anderson localisation is of the form

$$V(x) = \sum_{z \in \mathbb{Z}^d} \xi(z) v(x - z), \qquad x \in \mathbb{R}^d, \tag{1.18}$$

where $\xi = (\xi(z))_{z \in \mathbb{Z}^d}$ is a random i.i.d. field of random variables, and $v: \mathbb{R}^d \to [0, \infty)$ is a bounded, compactly supported cloud function, the so-called *single-site potential*. The popularity of this kind of potential in the community of Anderson localisation (see [Kir10] for an extended survey) comes from the fact that every $\mathbb{Z}^d$-ergodic potential V can be written in the form $V(x) = \sum_{z\in\mathbb{Z}^d} f_z(x-z)$ with suitable random variables f_z taking values in $L^p(\mathbb{R}^d)$, i.e., in a very similar way. However, to the best of our knowledge, the PAM with this kind of random potential has not yet been studied. ◇

Example 1.20 (Perturbed-lattice potential, also known as random displacement model) Another interesting choice is

$$V(x) = -\sum_{z\in\mathbb{Z}^d} W(x - z - \eta_z),$$

where $W:\mathbb{R}^d \to [0,\infty)$ is a single-site potential, and $(\eta_z)_{z\in\mathbb{Z}^d}$ is a sequence of centred $\mathbb{R}^d$-valued random variables. The interpretation is that, at each lattice site, a copy of W is intended to sit, but is randomly shifted by an individual amount. In the random Schrödinger operator community, this type of potentials is called the *random displacement model*. The PAM with this potential is analysed in [Fuk09a, FukUek10, FukUek11]. Most natural is to assume the η_z as i.i.d., but also just an ergodicity assumption is of interest. Let us remark that there is an interesting relation between the distribution of the set $\{z+\eta_z : z\in\mathbb{Z}^d\}$ (with $(\eta_z)_{z\in\mathbb{Z}^d}$ a particular ergodic sequence) and the set of zeros of a certain complex power series with i.i.d. Gaussian coefficients, see [Fuk09a, Remark 2]. ◇

Example 1.21 (Gaussian white noise) One of the most natural random potentials on $\mathbb{R}^d$ is an entirely uncorrelated one, the most natural example being a Gaussian white noise potential ξ, the centred Gaussian process $(V(x))_{x\in\mathbb{R}^d}$ with covariance $\langle V(x)V(y)\rangle = \delta_0(x-y)$ for $x,y\in\mathbb{R}^d$. Because of the low degree of regularity of such fields (actually, they are distributions instead of families of random variables), already the definition of the solution of the PAM presents a major challenge in $d\geq 2$, as it requires renormalisation procedures. The problem in a nutshell is that the regularity α of V is smaller than $-d/2$, and one expects a solution u of regularity $2+\alpha$, such that the product $u\cdot\xi$ should have a regularity $2+\alpha+\alpha$, which is negative. Hence, standard theory of stochastic partial differential equations does not apply. Furthermore, a meaningful construction is expected to be possible only in dimensions $d\leq 3$.

Only since very recently, techniques for overcoming these problems are being developed, and there is currently a high activity in that direction in the spirit of the theory of *rough paths* and novel methods like the theory of *regularity structures* [Hai13] and *paracontrolled distributions* [GubImkPer12]. Constructing solutions to the PAM is one of the few major application fields and test cases for these methods, the other prominent examples being the KPZ equation and the ϕ^4-model. The current state of the art is a construction almost surely on the entire state space $\mathbb{R}^d$; see [HaiLab15a] for $d = 2$ (even without usage of the heavy machinery of renormalisation procedures) and [HaiLab15b] for $d = 3$, however, with some restrictions with respect to the time-dependence.

The construction relies on a renormalisation with the help of a mollified version of V, i.e., its convolution with a smooth approximation of the delta measure with a parameter $\varepsilon > 0$ that has to be sent to zero. In order to obtain a (candidate for a) limit, one has to subtract a certain counter term from the equation that depends on ε and on the dimension, and the task is to prove that a limit of the solution to the modification exists. Due to the local character of the construction methods, the limit holds everywhere in space, but only on compact time intervals. Hence, it is not yet available in such a comfortable way that one could start thinking right away about intermittency questions.

Another natural way of construction of the solution would be in terms of a rescaling of the corresponding $\varepsilon\mathbb{Z}^d$-version with a certain ε-depending rescaling of an arbitrary i.i.d. random potential in the spirit of Donsker's invariance principle. Carrying out this way seems to be essentially within reach of the current state of the art, but has not yet been done; it is expected to be cumbersome.

Spectral theoretic questions for the Anderson Hamiltonian $\Delta + V$ with V a Gaussian white noise have been investigated in [AllCho15], however, only on the torus in $\mathbb{R}^2$ rather than the whole space. Using the concept of paracontrolled distributions introduced in [GubImkPer12], they give some sense to this operator as an unbounded self-adjoint operator on the space L^2 and show that its real spectrum is discrete. Furthermore, they approach this operator with a smoothed, renormalised version $\Delta + V_\varepsilon - c_\varepsilon$ as $\varepsilon \to 0$ with a suitable constant c_ε. Finally, they establish almost-sure asymptotics of the principal eigenvalue of $\Delta + V$ in a large torus of side length $L \to \infty$ on the scale $\log L$. $\diamond$

Chapter 2
Tools and Concepts

One of the most interesting features of the PAM is that, being a partial differential equation with random coefficients, it lies in the intersection of probability and functional analysis, which opens up exciting possibilities for combining tools from these two different parts of mathematics. Furthermore, there are classic and well-developed mathematical theories that enable explicit solution formulas and the application of further techniques to the study of the PAM. In this chapter, we give an account on these tools and pave the way for a deep understanding and a powerful analysis of the PAM. We bring the probabilistic side in Sect. 2.1 and the functional analytic side in Sect. 2.2. In Sect. 2.3, we discuss a number of aspects and conclusions that immediately follow from a combination of these tools; a panorama of precise conjectures arises.

2.1 Probabilistic Aspects

The PAM has a lot of relations to other questions and models, which explains the great interest that the PAM receives. We briefly survey the most important ones. Furthermore, we provide the most important probabilistic tools for the treatment of the PAM.

2.1.1 Branching Process with Random Branching Rates

The solution u to (1.1) admits an interpretation that arises from branching particle dynamics, see [GärMol90] and [CarMol94]. The following model is one important representative of a class of models called *branching random walk in random environment (BRWRE)*.

W. König, *The Parabolic Anderson Model*, Pathways in Mathematics,
DOI 10.1007/978-3-319-33596-4_2

Imagine that initially, at time $t = 0$, there is a single particle at the origin, and all other sites are vacant. This particle moves according to a continuous-time symmetric random walk with generator Δ^d. When present at site z, the particle splits into two particles with rate $\xi_+(z) \in (0, \infty)$ and is killed with rate $\xi_-(z) \in (0, \infty]$, where $\xi_+ = (\xi_+(z))_{z\in\mathbb{Z}^d}$ and $\xi_- = (\xi_-(z))_{z\in\mathbb{Z}^d}$ are independent random i.i.d. fields. Every particle continues from its birth site in the same way as the parent particle, and their movements are independent. Put $\xi(z) = \xi_+(z) - \xi_-(z) \in [-\infty, \infty)$. Then, given ξ_- and ξ_+, the expected number of particles present at the site z at time t, as a function of $(t, z) \in [0, \infty) \times \mathbb{Z}^d$, solves the equation (1.1) and is therefore, by uniqueness of the solution, equal to $u(t, z)$ [GärMol90]. Here the expectation is taken over the particle motion and over the splitting resp. killing mechanism, but not over the random medium (ξ_-, ξ_+). The fact that the expected particle number solves (1.1) is standard in the study of branching processes; see [Hol00] for an elementary derivation.

The successful work on the PAM since 1990 has fertilized also the study of the BRWRE, but to a surprisingly little extent yet. In Sect. 7.11 below, we survey some heuristics and results on the BRWRE that are influenced by the study of the PAM.

Remark 2.1 (Applications) The mathematical concept of spatial branching processes as described above has the following main applications.

- **Population dynamics.** It is a basic model for the spatial movement, branching and killing of indistinguishable particles in space. It may be seen as a drawback for realistic applications that the production of new particles at site does not depend on the current number of particles there, in particular the number of particles at a site is unbounded.
- **Mutation and selection.** If the space $\mathbb{Z}^d$ is replaced by a space of phenotypes of an individual, then the underlying branching process is a popular model for population genetics. Indeed, each jump within this space is interpreted as a mutation, as some detail in its biologic properties is changed, and the particles in the branching process are not registered according to the spatial location of the individuals, but according to their biological properties. One example is the replacement of $\mathbb{Z}^d$ by the N-dimensional hypercube $\{-1, 1\}^N$, which is a simple model space for the set of gene sequences that we mentioned in Remark 1.8.
- **Chemical reactions.** The particle model with migration, branching and killing also serves as a (very simple) model for chemical reactions. Indeed, imagine that particles are randomly distributed over $\mathbb{Z}^d$ that have an action as *catalysts* for a certain type of chemical reaction; that is, their presence at a given site supports the reaction of a certain *reactant* and helps producing new substance of it. In mathematical terms, we assume that a reactant particle at z is, for any catalyst particle present at z, split into two at a given rate $\gamma \in (0, \infty)$, say. That is, the rate of the reaction is linear in the number of catalysts. Additionally, assume that each reactant particle dies with fixed rate $\delta \in (0, \infty)$. Let $\xi_*(z)$ denote the

number of catalyst particles at z, whose presence we want to assume as random. Then $u(t, z)$ is the expected number of reactant particles at time t in the site z, where the random potential is given as $\xi = \gamma\xi_* - \delta$.

◇

Remark 2.2 (The (non-parabolic) Anderson equation) As we announced in Remark 1.1, there is an interpretation of the solution of the non-parabolic version of the PAM, the original quantum mechanic Schrödinger equation (1.4), in terms of a branching processes, see [Wag14, Wag15a, Wag15]. The underlying branching process is indeed a *marked* branching process with migration in $\mathbb{Z}^d$, and the marks are taken from the set $\{1, -1\} \times \{+, -\}$. While the marks 1 and -1 appear very natural to mark, in some way, the real part and the imaginary part, respectively, the introduction of the marks $+$ and $-$ are at the first sight surprising. They can be interpreted as 'present' and 'vanishing' or as 'visible' and 'hidden'. See [Wag14] for the precise mechanism. Then the solution to (1.4) is given as

$$u(t, z) = \mathbb{E}\big[\eta_{1,+}(t, z) - \eta_{1,-}(t, z)\big] + \mathrm{i}\,\mathbb{E}\big[\eta_{-1,+}(t, z) - \eta_{-1,-}(t, z)\big], \tag{2.1}$$

where $\eta_m(t, z)$ denotes the number of particles with mark m at time t in the site z. In [Wag15], a kind of Feynman-Kac formula is formulated in terms of a simple random walk on $\mathbb{Z}^d$. However, it seems more appropriate to formulate one for a suitable random walk on $\mathbb{Z}^d \times \{1, -1\} \times \{+, -\}$, but this is currently open. ◇

2.1.2 Feynman-Kac Formula

Some of the most interesting applications of the PAM are best explained in terms of an explicit formula for the solution in terms of random walks. A very useful standard tool for the probabilistic investigation of (1.1) is the well-known *Feynman-Kac formula* for the solution u, which reads

Feynman-Kac formula. Under the moment condition (1.5),

$$u(t, z) = \mathbb{E}_0\Big[\exp\Big\{\int_0^t \xi(X(s))\,\mathrm{d}s\Big\}\delta_z(X(t))\Big], \qquad (t, z) \in [0, \infty) \times \mathbb{Z}^d. \tag{2.2}$$

Here $(X(s))_{s\in[0,\infty)}$ is a continuous-time random walk on $\mathbb{Z}^d$ with generator Δ^{d} starting at $z \in \mathbb{Z}^d$ under $\mathbb{E}_z$. One can also write $\delta_z(X(t)) = \delta_{X(t)}(z) = \mathbb{1}\{X(t) = z\}$ for the indicator variable on the event that the endpoint of the path is located at z.

In words, in (2.2) a random walk path runs from the origin to z, and in the exponential we evaluate the sum of all the potential values that the walker sees on this way, weighted with the time that it spends in the respective lattice site. Actually, we used a time-reversal here, e.g., the initial condition δ_0 and the evaluation at z at time t may be interchanged. For u a solution with initial condition $u(0,\cdot) = u_0(\cdot)$, one would have to start the random walk at z and replace δ_z by u_0. By summing up over all $z \in \mathbb{Z}^d$, we see that the total mass $U(t)$ admits the Feynman-Kac representation

$$U(t) = \mathbb{E}_0\Big[\exp\Big\{\int_0^t \xi(X(s))\,\mathrm{d}s\Big\}\Big], \qquad t \in [0,\infty). \tag{2.3}$$

We refer the reader to [GärMol90, Theorem 2.1] for a proof of (2.2) (or its finite-space version in (2.7)), which is intimately connected with the almost sure existence and uniqueness of a solution to (1.1). Actually, the restriction to a finite box is technically much easier to handle (see Sect. 2.1.3); for a proof of finiteness of the infinite-space version (2.2) one has to control the decay of the potential ξ at infinity, and some percolation arguments are necessary for proving the uniqueness part, see our remarks after Theorem 1.2.

Remark 2.3 (Relation with semigroup theory) For the sake of better understanding, we give an explanation why the Feynman-Kac representation given in (2.2) actually solves problem (1.1) by making a connection to the theory of semi-groups of operators. Consider the family of operators $(\mathcal{P}_t)_{t\geq 0}$ acting on the bounded functions on the lattice as

$$\mathcal{P}_t f(z) = \mathbb{E}_z\Big[\exp\Big\{\int_0^t \xi(X(s))\,\mathrm{d}s\Big\} f(X(t))\Big], \qquad (t,z) \in [0,\infty) \times \mathbb{Z}^d. \tag{2.4}$$

By time reversal, we see that (2.2) is tantamount to $u(t,z) = \mathcal{P}_t\delta_0(z)$. An application of the Markov property shows that the family $(\mathcal{P}_t)_{t\geq 0}$ is a semi-group, i.e., $\mathcal{P}_0$ is the identical operator and $\mathcal{P}_s \circ \mathcal{P}_t = \mathcal{P}_{s+t}$ for any $s,t \in [0,\infty)$. Elementary calculations using the theory of continuous-time Markov chains reveal that the corresponding generator is equal to $\Delta^{\mathrm{d}} + \xi$ showing up on the right hand side of (1.1), i.e., $\partial_t \mathcal{P}_t f|_{t=0} = \Delta^{\mathrm{d}} f + \xi f$ for many functions f. (We do not enter here a discussion about the largest class of validity, i.e., a characterisation of the domain of the generator, but recall the criterion in (1.6).) Then we obtain the forward equation $\frac{\partial}{\partial t}\mathcal{P}_t f = (\Delta^{\mathrm{d}} + \xi)\mathcal{P}_t f$, which means that $u(t,z) = \mathcal{P}_t f(z)$ solves the parabolic Anderson problem (1.1) with initial condition $u(0,\cdot) = \mathcal{P}_0 f = f$. ◇

Remark 2.4 (Large-t asymptotics $\approx$ maximisation) By (2.2), $U(t)$ is the t-th exponential moment of the random variable $\frac{1}{t}\int_0^t \xi(X(s))\,\mathrm{d}s$. Hence, its large-$t$ asymptotics is intimately connected with the maximisation of this random variable, according to the well-known fact that the t-th exponential moment of a random variable behaves exponentially in t with rate equal to the essential supremum of that random variable. We will elaborate on this thought in Sect. 2.3.1. For the moment, we keep in mind that the large-t asymptotics of $U(t)$ are determined by paths X that make $\int_0^t \xi(X(s))\,\mathrm{d}s$ large. ◇

Remark 2.5 (Self-attractiveness) We see from the Feynman-Kac formula that the interaction with a random potential ξ induces a self-attractive effect to the path, when one takes the expectation with respect to ξ. Indeed, according to Remark 2.4, the paths that really count in the limit $t \to \infty$ are those who make the exponent $\int_0^t \xi(X(s))\,\mathrm{d}s$ large. When taking the expectation w.r.t. ξ, then the path and the potential jointly make $\int_0^t \xi(X(s))\,\mathrm{d}s$ large in a coordinated strategy, which consist of the following:

- ξ is very large in an area $B \subset \mathbb{Z}^d$, and
- the path does not leave B by time t.

Achieving large values is probabilistically costly, hence the potential would like to be large only in a small set B. On the other hand, not leaving a small set is probabilistically costly as well, so the path would like to do this only with a large set B. Potential and path have to find a compromise, i.e., an optimal size of B. This depends mainly on the costs for the potential to achieve extremely high values, i.e., on the upper tails of $\xi(0)$ (the asymptotics of $\mathrm{Prob}(\xi(0) > r)$ as $r \uparrow \operatorname{esssup} \xi(0)$). We will see in Chap. 3 how this compromise will be found, but it is clear from the central limit theorem that the diameter of an optimal set B will be much smaller than $\sqrt{t}$. ◇

It should be stressed that the random walk path in (2.2) is *not* to be interpreted as the trajectory along which the mass flows through the random potential, even though this association may be tempting. It does *not* reflect the time-evolution of the heat flow that is described by the solution $u(t,\cdot)$, but should be seen only as some mathematical object that enables an explicit description of $u(t,\cdot)$. Nevertheless, this path is often studied as an object on its own interest as some random path under the influence of the random environment ξ; see Sect. 2.1.5.

Certainly, there is a Feynman-Kac formula in the spatially continuous case as well, see many standard texts on Brownian motion, e.g., [Szn98, Section 1.1]. If $f\colon \mathbb{R}^d \to [-\infty,\infty)$ is a continuous function that is bounded from above, then the solution $u\colon (0,\infty)\times\mathbb{R}^d \to [0,\infty)$ of the PAM (1.7) (with V replaced by f) satisfies the formula

$$u(t,x) = \mathbb{E}_0\Big[\exp\Big\{\int_0^t f(Z(s))\,\mathrm{d}s\Big\}; Z(t)\in \mathrm{d}x\Big]\Big/\mathrm{d}x, \qquad t\in[0,\infty), \tag{2.5}$$

where $Z = (Z(s))_{s\in[0,\infty)}$ is a Brownian motion with generator Δ in $\mathbb{R}^d$ starting from x under $\mathbb{E}_x$. (Note that we dropped the factor $\frac{1}{2}$ in front of the Laplace operator, in accordance with (1.7).) This time, we need to conceive the expectation as a density of the terminal site $Z(t)$ in x, therefore we do not use the time-reversal property. In words, the right-hand side is the density of the random variable $Z(t)$ under the measure with density given by the exponential. The formula in (2.5) holds also under much weaker assumptions than upper boundedness of f [Szn98, Section 1.2];

it suffices to assume that $\limsup_{t\downarrow 0} \sup_{x\in\mathbb{R}^d} \mathbb{E}_x[\int_0^t |f(Z_s)|\,\mathrm{d}s] = 0$, i.e., that f lies in the *Kato class*.

Remark 2.6 (Renormalized Poisson potential) As we said in Example 1.15, one of the most natural and most-studied potentials is the Poisson trap potential $V(x) = -\int_{\mathbb{R}^d} W(x-y)\,\omega(\mathrm{d}y)$, where $\omega = \sum_i \delta_{x_i}$ is a standard Poisson point process in $\mathbb{R}^d$ with intensity $\nu \in (0,\infty)$, and $W\colon \mathbb{R}^d \to [0,\infty)$ is a continuous potential. However, for the choice $W(x) = C|x|^{-q}$ with $q \in (0,d]$, this potential is equal to $-\infty$ almost everywhere, almost surely [CheKul12, Proposition 2.1]. This type of potentials is worth being studied, since, in $d \geq 3$, for the choices $q = d-1$ and $q = d-2$ the potential has the interpretation of the gravitational force and potential, respectively. One way out of the problem is to consider the renormalised Poisson potential $V(x) = -\int_{\mathbb{R}^d} W(x-y)(\omega(\mathrm{d}y) - \mathrm{d}y)$. Indeed, if $\int_{\mathbb{R}^d}(\mathrm{e}^{-W(x)} - 1 + W(x))\,\mathrm{d}x < \infty$ (this is satisfied for $W(x) = C|x|^{-q}$ precisely in the case $q \in (d/2, d)$), then [CheKul12, Theorem 1.1] the renormalised potential can be properly defined, and [CheKul12, Proposition 1.2] the corresponding Feynman-Kac formula is a solution to (1.7). However, it is a solution possibly only in the weak sense, i.e., in the sense that

$$u(t,x) = u_0(x) + \int_0^t \mathrm{d}s \int_{\mathbb{R}^d} \mathrm{d}y\, p_{t-s}(x-y)u(s,y)\xi(y), \qquad x \in \mathbb{R}^d, t > 0,$$

and the integral on the right-hand side converges absolutely, where $p_s(x) = (4\pi s)^{-d/2}\mathrm{e}^{-|x|^2/4s}$ is the Gaussian density. In particular, the expectation of the Feynman-Kac formula (taken over the Poisson process) is finite, i.e., the first moment of the solution is finite. Let us also remark that one of the main formulas in [CheKul12] is

$$\langle U(t)\rangle = \Big\langle \mathbb{E}_0\Big[\exp\Big\{\int_0^t V(Z_s)\,\mathrm{d}s\Big\}\Big]\Big\rangle = \mathbb{E}_0\Big[\exp\Big\{\nu\int_{\mathbb{R}^d} F(w(t,x))\,\mathrm{d}x\Big\}\Big], \tag{2.6}$$

where $F(x) = \mathrm{e}^{W(x)} - 1 + W(x)$, $w(t,x) = \int_0^t W(Z_s - x)\,\mathrm{d}s$, and $(Z_s)_{s\in[0,\infty)}$ is a Brownian motion in $\mathbb{R}^d$ with generator Δ starting from x under $\mathbb{E}_x$. We will discuss the large-t behaviour of the moments in Sect. 7.3.4 below.

The first moment of the solution turns out to be infinite [CheKul12, Theorem 1.4] if we change the sign in front of the potential, i.e., if we consider $V(x) = C\int_{\mathbb{R}^d} |x-y|^{-q}(\omega(\mathrm{d}y) - \mathrm{d}y)$ with $C > 0$ and $q \in (d/2, d)$. However, it was shown in [CheKul12, Theorem 1.5] that the Feynman-Kac formula representing the solution to the PAM with this potential is almost surely finite for $q \in (d/2, \min\{2,d\})$, but infinite if $q > 2$. In the critical case $q = 2$ and $d = 3$ [CheRos11], the finiteness depends on whether $C < 1/16$ or not. The effect of another additional ingredient is studied in [CheXio15], where the Poisson process ω is assumed time-dependent, more precisely, it is replaced by the process $(\omega_s)_{s\in[0,\infty)}$ of independent Brownian motions, which is a Poisson point process at every time s. We refer to Sect. 7.3.4 for some asymptotic results for the moments for this model. ◇

2.1.3 Finite-Space Feynman-Kac Formulas

For many proofs, it will be important later to approximate the PAM with finite boxes. Luckily, the two most important types of boundary conditions turn out to serve for very useful lower and upper bounds, respectively.

If we equip the Anderson operator $\Delta^{\rm d} + \xi$ with zero boundary condition in some finite set $B \subset \mathbb{Z}^d$, then the corresponding solution u_B (see Remark 1.6) may be represented as

$$u_B(t, z) = \mathbb{E}_0\Big[\exp\Big\{\int_0^t \xi(X(s))\,{\rm d}s\Big\}\mathbb{1}\{X([0,t]) \subset B\}\mathbb{1}\{X(t) = z\}\Big], \tag{2.7}$$

i.e., the zero boundary condition is translated into the condition that the random walk does not leave B by time t. The Laplace operator with zero boundary condition in B generates the simple random walk before it exits B, i.e., restricted to not leaving B. It is clear that $u_B \leq u$ and that the total mass of u_B,

$$U_B(t) = \mathbb{E}_0\Big[\exp\Big\{\int_0^t \xi(X(s))\,{\rm d}s\Big\}\mathbb{1}\{X([0,t]) \subset B\}\Big], \tag{2.8}$$

satisfies $U_B \leq U$.

Now let $B = (-R, R]^d \cap \mathbb{Z}^d$ with $R \in \mathbb{N}$ be a centred box and consider the Anderson operator $\Delta^{\rm d} + \xi$ with periodic boundary condition in B. We obtain a Feynman-Kac formula by noting that the Laplace operator with periodic boundary condition generates the periodised simple random walk, $X^{(R)} = (X^{(R)}(s))_{s\in[0,\infty)}$, which can be pathwise realised as $X^{(R)}(s) = X(s) \bmod B$. This walk never leaves B. In plain words, if the walker is at the boundary of B and decides to jump to the outside of B, then it re-appears at the opposite side of B. Hence, we obtain

$$u_B^{\rm (per)}(t, z) = \mathbb{E}_0\Big[\exp\Big\{\int_0^t \xi(X^{(R)}(s))\,{\rm d}s\Big\}\mathbb{1}\{X^{(R)}(t) = z\}\Big], \tag{2.9}$$

and for its total mass:

$$U_B^{\rm (per)}(t) = \sum_{z\in B} u_B^{\rm (per)}(t, z) = \mathbb{E}_0\Big[\exp\Big\{\int_0^t \xi(X^{(R)}(s))\,{\rm d}s\Big\}\Big]. \tag{2.10}$$

We will see in Sect. 4.3 that, after taking expectation with respect to ξ, $U_B^{\rm (per)}(t)$ turns out to be an upper bound for $U(t)$, i.e., $\langle U(t)\rangle \leq \langle U_B^{\rm (per)}(t)\rangle$.

Remark 2.7 (One particle with random mass) The Feynman-Kac formulas in (2.2) and (2.7) and all variants will serve not only as starting points for several proofs, but also as settings for our intuition for the interpretation of the solution to the PAM (like also the branching process setting of Sect. 2.1.1). Indeed, we can now imagine that we start with one particle at the origin, carrying a unit mass at time zero. Then the particle starts its random walking along the trajectory in the Feynman-Kac formula and increases and decreases the mass that it carries according to the potential values that it sees on the way. At time t, its mass has the current value $\exp\{\int_0^t \xi(X_s)\,\mathrm{d}s\}$. Then $u(t,z)$ is its expectation, restricted to being at site z at that time. We will often refer to this picture and will talk about the 'random walker' or the 'trajectory of the Feynman-Kac formula'. When we talk about the 'optimal' one, then we mean those realisations of the trajectory that gives the best contribution to the expectation, i.e., maximising the value of $\exp\{\int_0^t \xi(X_s)\,\mathrm{d}s\}$ in comparison to the probability of that trajectory. ◇

2.1.4 Local Times and Moments

The functional in the exponent in the above Feynman-Kac formulas, $\int_0^t \xi(X(s))\,\mathrm{d}s$, is indeed a functional of the *local times* of the walk,

$$\ell_t(z) = \int_0^t \delta_z(X_s)\,\mathrm{d}s, \qquad t > 0,\, z \in \mathbb{Z}^d. \tag{2.11}$$

The family $(\ell_t(z))_{z\in\mathbb{Z}^d}$ is a random measure on $\mathbb{Z}^d$ with total mass equal to t. It registers the amount of time that the random walk spends in z up to time t. The *occupation times formula* says that

$$\int_0^t \xi(X_s)\,\mathrm{d}s = \sum_{z\in\mathbb{Z}^d} \xi(z)\ell_t(z). \tag{2.12}$$

Taking into account that the random potential ξ is i.i.d., we may easily calculate the expectation of the main term in the Feynman-Kac formula:

$$\begin{aligned}\langle \mathrm{e}^{\int_0^t \xi(X_s)\,\mathrm{d}s}\rangle &= \Big\langle \prod_{z\in\mathbb{Z}^d} \mathrm{e}^{\xi(z)\ell_t(z)}\Big\rangle = \prod_{z\in\mathbb{Z}^d} \langle \mathrm{e}^{\xi(0)\ell_t(z)}\rangle = \prod_{z\in\mathbb{Z}^d} \mathrm{e}^{H(\ell_t(z))}\\ &= \exp\Big\{\sum_{z\in\mathbb{Z}^d} H(\ell_t(z))\Big\},\end{aligned} \tag{2.13}$$

where we recall the logarithmic moment generating function $H(t) = \log\langle \mathrm{e}^{t\xi(0)}\rangle$ from (1.13). Certainly, for this calculation we have to assume that $H(t)$ is finite for all positive t, i.e., that all positive exponential moments of $\xi(0)$ are finite. Using Fubini's theorem for interchanging the two expectations, we arrive at

$$\langle U(t)\rangle = \mathbb{E}_0\Big[\exp\Big\{\sum_{z\in\mathbb{Z}^d} H(\ell_t(z))\Big\}\Big], \tag{2.14}$$

and similar formulas for the expectations of u, also for zero and periodic boundary conditions in some box B.

Remark 2.8 (Random motions in random media) We want to give a little guidance to the classification of the PAM within the world of random motions in random media; the vocabulary used in the probability literature has achieved a certain stability in this respect.

By the virtue of the Feynman-Kac formula in (2.2), the PAM is often called a *random walk in random potential*, and ξ is often called a *potential*. This is one of a handful of fundamental examples of a random motion in the presence of a random medium. Another one is the process $(\int_0^t \xi(X(s))\,\mathrm{d}s)_{t\in[0,\infty)}$ that appears in the exponent of the Feynman-Kac formula, which is sometimes called the *random walk in random scenery (RWRSc)*. This is an interesting object to study on its own, also in discrete time and for Brownian motion instead of random walks. In recent years, several authors got interested in the description of its extreme behaviour, which, on a technical level, has much to do with the analysis of the PAM, see Sect. 7.4. A third example is the *random walk in random environment (RWRE)*, whose transition probabilities are given by a random field of probability measures in the sites of $\mathbb{Z}^d$. Conditional on the environment (i.e., in the quenched setting), is a Markov process, but not in the annealed setting, i.e., when the environment is averaged out. Important examples are the *random walk among random conductances (RWRC)*, which we consider in Sect. 7.9, and the *Bouchaud random walk*, see Sect. 7.9.2. To complete this small list (without considering them further), also *self-interacting random walks* are fundamental, which evolve in time according to random mechanisms depending on the past, often only on the local times produced so far. Important examples here are the *reinforced random walk* and the *myopic random walk* or *true self-repellent random walk*. $\diamond$

2.1.5 Quenched and Annealed Transformed Path Measures

Starting from the Feynman-Kac formula (2.2), it is rather natural to consider the *quenched path measure*

$$Q_{\xi,t}(\mathrm{d}X) = \frac{\mathrm{e}^{\int_0^t \xi(X(s))\,\mathrm{d}s}}{U(t)}\mathbb{P}_0(\mathrm{d}X), \qquad t \geq 0, \tag{2.15}$$

on the set of trajectories $[0,t] \to \mathbb{Z}^d$, $[0,t] \ni s \mapsto X(s)$. Note that $Q_{\xi,t}$ is a probability measure by (2.2). It obviously depends on the realisation of the potential ξ and does in general not constitute a consistent family of measures in t, i.e., they do not come from a path measure for paths $[0,\infty) \to \mathbb{Z}^d$ by projection on the time interval $[0,t]$.

Like in many models of statistical mechanics, the study of the large-t asymptotics of $U(t)$ is intimately connected with the question of the behaviour of $(X(s))_{s\in[0,t]}$ under $Q_{\xi,t}$. In particular, we will often discuss intermittency (see Sect. 1.4) in this view. Plainly, intermittency is reflected by the behaviour of the random walker X under $Q_{\xi,t}$ to move quickly through the potential landscape to reach some region(s) of exceptionally high potential and then stay there up to time t. This would make the integral in the numerator on the right of (2.15) especially large, and it would give much weight to trajectories that end at time t in such regions. Certainly, these regions are the intermittent islands, and it may a priori be that different trajectories choose different islands. On the other hand, the probability (under $\mathbb{P}_0$) to quickly reach such a distant potential peak may be rather small, since the optimal regions are typically far away. Hence, the main mass in $Q_{\xi,t}$ comes from paths that find a good compromise between the high potential values and the far distance, and so does the main contribution to $U(t)$. This contribution is mainly given by the absolute height of the peak. The second-order contribution to $U(t)$ is determined by some finer information, for example, by geometric properties of the potential in that peak.

In analogy, the *annealed path measures* are defined as

$$Q_t(\mathrm{d}X) = \frac{\langle \mathrm{e}^{\int_0^t \xi(X_s)\,\mathrm{d}s}\rangle}{\langle U(t)\rangle}\,\mathbb{P}_0(\mathrm{d}X) = \frac{\mathrm{e}^{\sum_{z\in\mathbb{Z}^d} H(\ell_t(z))}}{\mathbb{E}_0[\mathrm{e}^{\sum_{z\in\mathbb{Z}^d} H(\ell_t(z))}]}\,\mathbb{P}_0(\mathrm{d}X), \qquad t \geq 0, \qquad (2.16)$$

where we recall the local times and the cumulant generating function from Sect. 2.1.4. It might be confusing that the density in (2.16) is *not* chosen as the expectation of the density in (2.15), which would also make perfect sense as an annealed path measure (note that the term 'annealed' is not a mathematical term, but depends on the taste and the view of the author). However, the main objective is the analysis of the partition function, i.e., the term $\mathrm{e}^{\int_0^t \xi(X_s)\,\mathrm{d}s}$, and therefore the impact of its expectation is of principal interest.

In (4.5) below we will see that the density in (2.16) has an attractive effect on the path, as the functional $\mu \mapsto \exp\{\sum_z H(t\mu(z))\}$, seen as a map on probability measures on $\mathbb{Z}^d$, is convex. Hence, one may already here expect that the walk will, under Q_t, spread out on a smaller area than the free random walk, i.e., we may expect that $X_t \ll \sqrt{t}$ as $t \to \infty$, typically under Q_t. See Sect. 7.5 for results on upper tails for the functional in the exponent and Sect. 7.6 for results on such path measures.

Example 2.9 (Simple random walk among Bernoulli traps) The simple case where ξ is an i.i.d. field and each potential value is either $-\infty$ or 0 (see Example 1.10) is called *simple random walk among Bernoulli traps*. The solution to the PAM may be seen as the survival probability of the walk. Indeed, let

$$\mathcal{O} = \{z \in \mathbb{Z}^d : \xi(z) = -\infty\}$$

be the set of obstacles or traps, then it is clear that the exponent $\int_0^t \xi(X(s))\,\mathrm{d}s$ in the Feynman-Kac formula is equal to $-\infty$ as soon as the path $X([0,t])$ hits $\mathcal{O}$. This implies that

$$u(t,z) = \mathbb{P}_0(X([0,t]) \subset \mathcal{O}^{\mathrm{c}}, X(t) = z)$$

is the probability that the path does not hit any trap by time t and ends up at the site z, and $U(t)$ is the survival probability. Introducing the stopping time $T_{\mathcal{O}} = \inf\{t > 0 : X(t) \in \mathcal{O}\}$ of the first visit to the obstacles, we may also write $u(t,z) = \mathbb{P}_0(T_{\mathcal{O}} > t, X(t) = z)$, and $U(t) = \mathbb{P}_0(T_{\mathcal{O}} > t)$ is the upper tail of $T_{\mathcal{O}}$. Hence, the measure $Q_{\xi,t}$ defined in (2.15) has the density $\mathbb{1}\{T_{\mathcal{O}} > t\}/\mathbb{P}_0(T_{\mathcal{O}} > t)$.

The density of the annealed measure, Q_t, can easily be calculated from (2.13), since

$$\sum_{z\in\mathbb{Z}^d} H(\ell_t(z)) = (\log p) \sum_{z\in\mathbb{Z}^d} \mathbb{1}\{\ell_t(z) > 0\} = R_t \log p,$$

where $R_t = |\{X(s): s \in [0,t]\}|$ is the *range* of the walk by time t, the number of visited sites. Hence, the density of Q_t with respect to the simple random walk measure is equal to $\mathrm{e}^{-\nu R_t}/\mathbb{E}_0[\mathrm{e}^{-\nu R_t}]$, where $\nu = -\log p \in (0,\infty)$. That is, the expected total mass of the solution of the PAM, $\langle U(t)\rangle = \mathbb{E}_0[\mathrm{e}^{-\nu R_t}]$, is equal to a negative exponential moment of the range. The large-t study of the latter has been called the *range problem*.

The intermittent islands are the ones where $u(t,\cdot)$ achieves its maximum zero, the trap-free zones. It will turn out that these islands depend on t and are rather large; in fact, in the annealed setting their radius is of order $t^{1/(d+2)}$, and in the quenched setting they are of order $(\log t)^{1/(d+2)}$.

Let us mention that a discussion of general trapping problems from a physicist's and a chemist's point of view, including a survey on related mathematical models and a collection of open problems, is provided in [HolWei94]. ◇

Example 2.10 (Brownian motion among Poisson traps) Recall the trapped Brownian motion of Example 1.15 and look at the special case $V(x) = -\infty \times \sum_i \mathbb{1}\{x \in K_a(x_i)\}$, where we use $K_a(x)$ to denote the ball with radius a around x, and $(x_i)_i$ is a Poisson point process in $\mathbb{R}^d$ with intensity $\nu \in (0,\infty)$. Hence, $V(x)$ is equal to $-\infty$ in the a-neighbourhood of the union of the Poisson points. The solution u to the PAM is also known under the name *Brownian motion among Poisson traps* or *Brownian motion among Poisson obstacles*, as it is equal to a survival probability. Indeed, let $\mathcal{O} = \bigcup_i K_a(x_i)$ be the union of the a-balls around the Poisson points, then $V(x) = -\infty\mathbb{1}\{x \in \mathcal{O}\}$. Consider the stopping time $T_{\mathcal{O}} = \inf\{t > 0: Z_t \in \mathcal{O}\}$, the first entry time of the Brownian motion $(Z(s))_{s\in[0,\infty)}$ into the obstacle set $\mathcal{O}$, then we have $\int_0^t V(Z(s))\,\mathrm{d}s = -\infty\mathbb{1}\{T_{\mathcal{O}} \leq t\}$. The Feynman-Kac representation reads

$$u(t,x) = \mathbb{P}_0\left(T_{\mathcal{O}} > t,\, Z(t) \in \mathrm{d}x\right)/\mathrm{d}x,$$

i.e., $u(t,x)$ is equal to the sub-probability density of $Z(t)$ on survival in the Poisson field of traps by time t. The total mass $U(t) = \mathbb{P}_0(T_{\mathcal{O}} > t)$ is the survival probability by time t. The analogue of the path measure $Q_{\xi,t}$ is the conditional distribution given the event $\{T_{\mathcal{O}} > t\}$, i.e., it transforms with the Radon-Nikodym density $\mathbb{1}\{T_{\mathcal{O}} > t\}/U(t)$.

It is easily seen that the first moment of $U(t)$ coincides with a negative exponential moment of the volume of the *Wiener sausage* $S_a(t) = \bigcup_{s\in[0,t]} K_a(Z(s))$, i.e.,

$$\begin{aligned}\langle U(t)\rangle &= \mathbb{E}_0\big[\langle \mathbb{1}\{Z([0,t]) \cap \mathcal{O} = \emptyset\}\rangle\big] = \mathbb{E}_0\big[\langle \mathbb{1}\{\#\{i: x_i \in S_a(t)\} = 0\}\big] \\ &= \mathbb{E}_0[\mathrm{e}^{-\nu|S_a(t)|}],\end{aligned} \tag{2.17}$$

where $\nu \in (0,\infty)$ is the intensity of the Poisson process, and $|\cdot|$ denotes the Lebesgue measure. For this reason, the analysis of the annealed transformed path measure Q_t is sometimes called the *Wiener sausage problem*; it was historically the first special case of a PAM for which substantial asymptotic results were derived [DonVar75]; see Sect. 3.5.1. ◇

2.2 Functional Analytic Aspects

It belongs to the standard knowledge of functional analysis that the solution to the heat equation with potential ξ in a finite box can be represented in terms of an *eigenvalue expansion* (also called *Fourier expansion*), i.e., an expansion with respect to the spectrum of the operator on the right-hand side of (1.1), the *Anderson Hamiltonian* $\Delta^{\mathrm{d}}+\xi$. This is one of the most important and fruitful connections of the heat equation with analytic theory. In this section we introduce the relevant notions and recall the most important facts. Recall that we do not put a minus sign in front of the Laplace operator, unlike the mathematical physics community. In particular, we do not speak of the 'bottom of the spectrum' but of the 'top', and 'deep valleys' of the potential are here 'high exceedances'.

2.2.1 *Eigenvalue Expansion*

We introduce Dirichlet (i.e., zero) boundary condition in a finite set $B \subset \mathbb{Z}^d$ and denote the Hamilton operator $\Delta^{\mathrm{d}} + \xi$ by $\mathcal{H}_B$. This operator is symmetric and non-negative definite on the Hilbert space $\ell^2(B)$ of square-integrable functions (vectors) on B. Furthermore, it has precisely $|B|$ eigenvalues $\lambda_1(B) > \lambda_2(B) \geq \lambda_3(B) \geq \cdots \geq \lambda_{|B|}(B)$, and there is an orthonormal basis of $\ell^2(B)$ consisting of corresponding eigenfunctions (eigenvectors) $v_1, v_2, v_3, \ldots, v_{|B|}$, which also depend certainly on B. We always take the *principal eigenfunction* v_1 positive everywhere in B, while all the other eigenvectors may have positive and negative values in B. We think of v_k as being defined on the entire space $\mathbb{Z}^d$ with $v_k(z) = 0$ in B^{c}. Certainly, all these eigenvalues and eigenvectors are random, as they depend on ξ, but for a while this will not be important.

Now we consider the solution u_B of (1.1) in B with localised initial condition $u_B(0,\cdot) = \delta_0(\cdot)$; see Remark 1.6. It admits the spectral representation (sometimes also called *Fourier expansion* or *spectral decomposition*)

$$u_B(t,\cdot) = \sum_{k=1}^{|B|} \mathrm{e}^{t\lambda_k(B)} v_k(0) v_k(\cdot), \qquad t \in (0,\infty), \tag{2.18}$$

with respect to the eigenvalues and eigenfunctions. This can also be written as

$$u_B(t,z) = \langle \delta_z, \mathrm{e}^{t\mathcal{H}_B}\delta_0\rangle = \big(\mathrm{e}^{t\mathcal{H}_B}\delta_0\big)(z), \tag{2.19}$$

where $\langle\cdot,\cdot\rangle$ denotes the standard inner product in $\ell^2(\mathbb{Z}^d)$. The δ_0 is the initial condition, the δ_z refers to the evaluation at z, which can also be seen as a terminal condition.

One of the most obvious nice things about (2.18) is that the time-dependence sits exclusively in the exponent as a prefactor of the eigenvalues. In particular, one immediately sees that the main part of the large-t behaviour of $u_B(t,\cdot)$ should come from the principal eigenvalue, $\lambda_1(B)$. This fact is not drastically altered if the box B depends on t and grows to $\mathbb{Z}^d$ for large t, but will need some more care; actually there are cases in which the main contribution does not come from the first eigenvalue, but from another eigenvalue λ_k that has a better value of $v_k(0)$.

Some drawback about (2.18) is that there is a priori no version for $B = \mathbb{Z}^d$, at least not for random i.i.d. potentials ξ, unlike the Feynman-Kac formula. Versions of (2.18) on the entire space $\mathbb{Z}^d$ require that the potential decreases to $-\infty$ far out (i.e., $\lim_{|z|\to\infty}\xi(z) = -\infty$), as this implies that the Hamiltonian $\Delta^{\mathrm{d}} + \xi$ has a compact resolvent on $\ell^2(\mathbb{Z}^d)$, but this is not satisfied for an i.i.d. potential ξ, unlike in trivial cases. I am not aware of versions of (2.18) on $\mathbb{Z}^d$ that work under conditions like (1.5), but possibly they would not be too helpful.

2.2.2 *Relation Between Eigenvalue Expansion and the PAM*

The eigenvalue expansion in (2.18) yields an instructive explanation of the large-t asymptotics from a spectral point of view and serves as a starting point for powerful proofs, see also Sect. 2.2.3. Let us illustrate some the of the benefits for the study of the PAM that (2.18) offers.

Rayleigh-Ritz formula. One certainly guesses that the large-t asymptotics of the function $u_B(t,\cdot)$ should be mainly governed by the principal eigenvalue, $\lambda_1 = \lambda_1(B)$, and this is true for many levels of precision. Therefore, the *Rayleigh-Ritz formula* or *Rayleigh-Ritz principle* is of high interest:

$$\begin{aligned}\lambda_1(B) &= \sup_{v\in\ell^2(\mathbb{Z}^d):\mathrm{supp}\,(v)\subset B,\|v\|_2=1} \langle \mathcal{H}_B v, v\rangle \\ &= -\inf_{v\in\ell^2(\mathbb{Z}^d):\mathrm{supp}\,(v)\subset B,\|v\|_2=1} \Big(\frac{1}{2}\sum_{x,y\in\mathbb{Z}^d:x\sim y}(v_x-v_y)^2-\sum_{z\in B}\xi(z)v_z^2\Big),\end{aligned} \tag{2.20}$$

where we wrote $x \sim y$ to denote that x and y are nearest neighbours, i.e., they differ in precisely one component and precisely by one. (In the sum on x, y we mean the sum on the ordered pairs (x, y), i.e., $(x, y) \neq (y, x)$ for $x \neq y$, which gives rise to the prefactor of $\frac{1}{2}$.) We remark that the first sum in the second line can be restricted to the sum on x and y in B and its outer boundary. The variational formula on the right-hand side has precisely one solution up to a multiplicative constant, and this is v_1, which is a positive vector.

Upper estimates for u_B. There is a standard way to estimate the total mass of u_B in terms of the principal eigenvalue with the help of the Cauchy-Schwarz inequality ($\langle f, gk\rangle \leq \|f\|_2\,\|g\|_2$ for any $f, g \in \ell^2(B)$) and Parseval's identity ($\sum_{k=1}^{|B|}\langle v_k, f\rangle^2 = \|f\|_2^2$ for any $f \in \ell^2(B)$) as follows:

$$\begin{aligned}U_B(t) = \sum_{z\in B} u_B(t,z) &= \sum_{k=1}^{|B|} \mathrm{e}^{t\lambda_k}\langle v_k,\delta_0\rangle\langle v_k,\mathbb{1}\rangle \\ &\leq \Big(\sum_{k=1}^{|B|}\mathrm{e}^{t\lambda_k}\langle v_k,\delta_0\rangle^2\Big)^{1/2}\Big(\sum_{k=1}^{|B|}\mathrm{e}^{t\lambda_k}\langle v_k,\mathbb{1}\rangle^2\Big)^{1/2} \\ &\leq \mathrm{e}^{t\lambda_1}\Big(\sum_{k=1}^{|B|}\langle v_k,\delta_0\rangle^2\Big)^{1/2}\Big(\sum_{k=1}^{|B|}\langle v_k,\mathbb{1}\rangle^2\Big)^{1/2} \\ &= \mathrm{e}^{t\lambda_1}\|\delta_0\|_2\|\mathbb{1}\|_2 = \mathrm{e}^{t\lambda_1}\sqrt{|B|}.\end{aligned} \tag{2.21}$$

See Remark 3.1 for the approximation of U with U_B for large boxes B.

Lower estimates for u_B. In in the investigation of the PAM, it turned out useful to reverse the estimate in (2.21), i.e., to estimate the eigenvalue λ_1 in terms of the solution u_B, with the help of the expansion in (2.18). This seems difficult on the first sight, since all eigenfunctions v_k, with the exception of v_1, assume positive and negative values. However, if one plays with the initial condition, this problem is removed. So let us denote by $u_B^{(y)}$ the solution to (1.1) with initial condition

$u_B^{(y)}(0,\cdot) = \delta_y(\cdot)$ instead of $\delta_0(\cdot)$, then we can estimate, using that every v_k is ℓ^2-normalised,

$$\mathrm{e}^{t\lambda_1} \leq \sum_{k=1}^{|B|} \mathrm{e}^{t\lambda_k} = \sum_{k=1}^{|B|} \mathrm{e}^{t\lambda_k} \sum_{x\in B} v_k(x)^2 = \sum_{x\in B}\sum_{k=1}^{|B|} \mathrm{e}^{t\lambda_k} \langle v_k, \delta_x\rangle^2 = \sum_{x\in B} u_B^{(x)}(t,x). \tag{2.22}$$

Applying the Feynman-Kac formula in (2.7), adapted to the initial condition δ_y, we arrive at expressions that can be handled further with the same means as $U_B(t) = \sum_{x\in B} u_B(t,x)$, and they give the same leading asymptotics.

2.2.3 Anderson Localisation

One of the great sources of interest in the random Schrödinger operator $\Delta^{\mathrm{d}} + \xi$ is the fact that its spectral properties help describing electrical conductance properties of alloys of metals or optical properties of glasses with random impurities. Therefore, one is naturally interested in *bounded* potentials, as the potential generically models the concentration ratio of the two metals in the conductance application.

Another source of interest in the spectrum of $\Delta^{\mathrm{d}} + \xi$ is the study of the Anderson Schrödinger equation in (1.4), which differs from the PAM by the additional prefactor of the imaginary unit i on the left-hand side of (1.1) and models dynamical quantum mechanical processes in a random environment. Note that in the large-t analysis of this equation, the entire spectrum is involved, in contrast to the large-t asymptotics of the PAM, where only the top of the spectrum is involved.

The driving force for studying the spectral properties of $\Delta^{\mathrm{d}} + \xi$ comes from the exciting prediction of P.W. Anderson [And58] that the spectrum should have a peculiar behaviour, which in a way interpolates between the smoothing effect of the Laplace operator, whose 'eigenfunctions' are infinitely spread out in $\mathbb{Z}^d$, and the localising effect of the multiplication operator ξ, which has only delta functions as eigenfunctions. He predicted that, at least in its spectrum close to the spectral ends (we are thinking of a bounded random potential ξ, for which also the spectrum of $\Delta^{\mathrm{d}} + \xi$ is bounded), all eigenfunctions of $\Delta^{\mathrm{d}} + \xi$ should be exponentially localised. More precisely, for all eigenvalues close to any of the two boundaries of the spectrum, the corresponding eigenfunction should decay exponentially fast away from its individual (random) localisation centre. This predicted phenomenon is nowadays called *Anderson localisation*. It was the motivation of an intense research activity in the last decades, and its validity has meanwhile been confirmed in a great number of cases, after the invention of deep mathematical tools. See [Kir10] for an extensive, and pedagogically written, survey on Anderson localisation and further reading.

2.2.4 Intermittency and Anderson Localisation

Let us explain how Anderson localisation is related with intermittency in the PAM. The starting point is the spectral representation in (2.18) with a large box B (depending on t) such that u_B is a good approximation for u (see Remark 3.1). In the limit $t \to \infty$, we can neglect all the summands in (2.18) with large k, because the exponential term $e^{t\lambda_k}$ makes them negligible in comparison to the leading terms $e^{t\lambda_1}, e^{t\lambda_2}, e^{t\lambda_3}, \dots$. According to the Anderson localisation prediction, at least for small k, the eigenfunctions v_k should be exponentially localised in centres x_k. Here we anticipate that the localisation property, which is predicted by Anderson localisation theory only in the entire space $\mathbb{Z}^d$, persists to large boxes. Moreover, as extreme-value statistics predicts (see Sect. 6.4 below), these centres are far away from each other, since they form a Poisson point process, after rescaling (see Sect. 6.3 below). Hence, v_k should be small outside a finite neighbourhood of x_k and even extremely small in neighbourhoods of the other x_is and in the origin. Hence, $u_B(t, x_k + \cdot)$ is well-approximated in a neighbourhood of zero by just the kth term, $e^{t\lambda_k} v_k(0) v_k(x_k + \cdot)$. As a consequence, the field $u_B(t, \cdot)$ has high peaks in small islands (the neighbourhoods of the localisation centres of the leading eigenvalues), which are far away from each other, and is much smaller outside these islands. This is a clear picture of intermittency. Additionally, we also see that the solution $u(t, \cdot)$ should be shaped like the eigenfunctions in these islands.

2.2.5 Integrated Density of States

We saw in Sect. 2.2.4 that large-t asymptotics of the PAM have much to do with the top of the spectrum (eigenvalues and the corresponding eigenfunctions) of the Anderson Hamiltonian $\Delta^{\mathrm{d}} + \xi$ with zero boundary condition in large boxes. Another explicit manifestation of this relation is in terms of *Lifshitz tails*, which describe the upper tails of the *integrated density of states (IDS)*.

One definition of the integrated density of states is as follows, see [CarLac90, Kir10]. In order to be consistent with the literature, we consider the operator $-\Delta^{\mathrm{d}} - \xi$. By $(-\Delta^{\mathrm{d}} - \xi)_{B_R}$ we denote its restriction to the box $B_R = [-R, R] \cap \mathbb{Z}^d$ with zero boundary condition. Denote by $E_1 < E_2 \leq E_2 \leq \dots \leq E_{|B_R|}$ its eigenvalues, counted with multiplicity (and of course depending on R). Let $N_R = \sum_k \delta_{E_k}$ denote its spectral measure. For an energy $E \in \mathbb{R}$, let

$$\mu_R(E) = N_R((-\infty, E]) \tag{2.23}$$

denote the number of eigenvalues $\leq E$ of $(-\Delta^{\mathrm{d}} - \xi)_{B_R}$. Then, by the subadditive ergodic theorem, the limit

$$\mu(E) = \lim_{R \to \infty} \frac{1}{|B_R|} \mu_R(E) \tag{2.24}$$

exists and is almost surely constant. The function μ is called the *integrated density of states (IDS)*. The interpretation of $\mu(E)$ is the number of energy levels of $-\Delta^{\rm d}-\xi$ below E per unit volume. Note that $\mu(E) \in [0,1]$, since the $B_R \times B_R$-matrix $(-\Delta^{\rm d} - \xi)_{B_R}$ cannot have more eigenvalues than the cardinality of B_R. After shifting and rescaling, μ is a distribution function, i.e., it is increasing and right-continuous with left limits and boundary values 1 as $E \uparrow \sup\sigma(-\Delta^{\rm d}-\xi)$ and 0 as $E \downarrow \inf\sigma(-\Delta^{\rm d}-\xi)$, where $\sigma(\mathcal{H})$ denotes the spectrum of an operator $\mathcal{H}$.

The IDS is related to the PAM as follows. Let

$$\mathcal{L}(N_R, t) = \int_{\mathbb{R}} {\rm e}^{-\lambda t}\, N_R({\rm d}\lambda) = \sum_k {\rm e}^{-tE_k} \tag{2.25}$$

be the Laplace transform of N_R evaluated at $t > 0$. Using the eigenvalue expansion in (2.18) and the finite-box Feynman-Kac formula in (2.7), we have the representation

$$\mathcal{L}(N_R, t) = \sum_{z\in B_R} \mathbb{E}_z\Big[{\rm e}^{\int_0^t \xi(X_s)\,{\rm d}s} \mathbb{1}\{X_{[0,t]} \subset B_R\}\mathbb{1}\{X_t = z\}\Big], \tag{2.26}$$

i.e., the sum over $z \in B_R$ of solutions to the PAM with initial condition δ_z, evaluated at z; see (2.22). The existence of the limit in (2.24) is proved by showing that $\frac{1}{|B_R|}N_R$ has an almost sure limit N, and this in turn is proved by showing that $\frac{1}{|B_R|}\mathcal{L}(N_R, t)$ has a non-trivial limit for any t. Using the ergodic theorem in (2.26), it is not difficult to prove that, almost surely,

$$\lim_{R\to\infty} \frac{1}{|B_R|}\mathcal{L}(N_R, t) = \Big\langle \mathbb{E}_0\Big[{\rm e}^{\int_0^t \xi(X_s)\,{\rm d}s}\mathbb{1}\{X_t = 0\}\Big]\Big\rangle = \langle u(t,0)\rangle. \tag{2.27}$$

Hence, $\frac{1}{|B_R|}N_R$ has a limit N as $R \to \infty$, whose Laplace transform $\mathcal{L}(N, t)$ is given by the right-hand side of (2.27), and this is equal to the expectation of the solution to the PAM as in (1.1) evaluated at zero. Certainly, $\mu(E) = N((-\infty, E])$ is the distribution function of N.

There is also a useful connection between the IDS and the principal eigenvalue in a fixed box [CarLac90, VI.15, p. 311]. Indeed, for any $R \in \mathbb{N}$,

$$\mu(E) \geq \frac{1}{|B_R|}\langle \mu_R(E)\rangle \geq \frac{1}{|B_R|}\sum_k {\rm Prob}(E_k \leq E) \geq \frac{1}{|B_R|}{\rm Prob}(E_1 \leq E). \tag{2.28}$$

This connection was utilized in [Fuk09b] for deriving relations between asymptotics of $\mu(E)$ for $E \downarrow \inf\sigma(-\Delta^{\rm d} - \xi)$ (see Sect. 2.2.6) and the almost sure asymptotics of the principal eigenvalue in large boxes.

2.2.6 Lifshitz Tails

Roughly speaking, the logarithmic asymptotics of the IDS $\mu(E)$ (as defined in (2.24) above) for $E \downarrow \inf \sigma(-\Delta^{\mathrm{d}} - \xi)$ are called the *Lifshitz tails* of the operator $-\Delta^{\mathrm{d}} - \xi$, see [CarLac90, Kir10]. They are of high interest for the description of the spectrum of $-\Delta^{\mathrm{d}} - \xi$ close to its bottom. But they are intimately connected with the large-t asymptotics of the PAM by the fact that they stand in a one-to-one connection with its Laplace transform $\mathcal{L}(N, t)$ of the IDS N. According to (2.27), we have $\mathcal{L}(N, t) = \langle u(t, 0)\rangle$, where u is the solution to the PAM with initial condition δ_0.

In this book we are mostly concerned with moment asymptotics for the total mass (rather than $u(t, 0)$), in particular in Chap. 3. However, it is not difficult (see below) to see that

$$\mathcal{L}(N, t) \approx \langle U(t)\rangle \qquad \text{as } t \to \infty, \tag{2.29}$$

where we mean by '$\approx$' logarithmic equivalence, i.e., the quotient of the logarithms of the two sides converges to one. Based on (2.29), one easily derives Lifshitz tails from large-t moment asymptotics of the total mass. We refer to Remark 3.21 for an explicit example of such assertion that has been proved in the literature.

A proof of (2.29) is not very difficult. Indeed, '$\leq$' is obvious from the Feynman-Kac formula in (2.2) (just drop the indicator on $\{X(t) = 0\}$), and one obtains a lower bound for $\mathcal{L}(N, t)$ by inserting, on the right-hand side of (2.27), the indicator on $\{X([0, t]) \subset B\}$ for any set B. A good choice for B is a t-dependent large centred box; see Remark 3.1. Expanding this in an eigenvalue series, we obtain

$$\mathcal{L}(N, t) \geq \Big\langle \sum_k \mathrm{e}^{t\lambda_k(B)} v_k(0)^2 \Big\rangle \geq \big\langle \mathrm{e}^{t\lambda_1(B)} v_1(0)^2 \big\rangle.$$

(Note that it was the coincidence of the initial and terminal conditions that enabled us to drop all other summands.) Now some technical work is required to deduce that the term $v_1(0)^2$ is negligible. The fact that $\langle \mathrm{e}^{t\lambda_1(B)}\rangle \approx \langle U(t)\rangle$ can be proved starting from the estimate in (2.21) and using Remark 3.1.

2.3 First Heuristic Observations

Based on the probabilistic and the functional analytic considerations in Sects. 2.1 and 2.2, let us give now some heuristics about what to expect in the description of the solution of the PAM in the long-time limit.

2.3.1 The Total Mass as an Exponential Moment

The first observation is that, via the Feynman-Kac formula in (2.3), $U(t)$ is equal to the t-th positive exponential moment of the quantity

$$Y_t = \frac{1}{t}\int_0^t \xi(X_s)\,\mathrm{d}s,$$

the average of the potential values along the random walk path. (The quantity tY_t is sometimes called *random walk in a random scenery*, see Remark 2.8.) It is a well-known fact from standard probability theory that, for any random variable Y, we have $\lim_{t\to\infty}\frac{1}{t}\log\mathbb{E}[\mathrm{e}^{tY}] = \operatorname{esssup} Y \in (-\infty,\infty]$. Hence, the limiting exponential growth rate of $U(t)$ as $t \to \infty$ will have much to do with the maximisation of Y_t over the probability space.

Actually, this maximisation has to be put into the right balance with the limiting behaviour of Y_t as $t \to \infty$, i.e., with the prefactor t in the exponent. If one considers the expectation of $U(t)$ with respect to ξ, one has to find a balance between the two random objects, the path and the potential. Certainly, an optimisation of Y_t is achieved by confining the random walk path $(X(s))_{s\in[0,t]}$ to an area in which the potential ξ is extremely large, and in which it does not cost the path too much to stay a long time. This area will be centred around the starting point of the motion, and it will be much smaller than of the size that one knows from the central limit theorem, i.e., its diameter will be much smaller than $\sqrt{t}$. On the other hand, picking just one site in which the potential is extremely huge will not necessarily be optimal.

Hence, the upper tails of ξ (i.e., the asymptotics of $\mathrm{Prob}(\xi(0) > r)$ for $r \uparrow \operatorname{esssup}\xi(0)$) will be one of the most decisive criteria, since they quantify the probabilistic cost of making the potential large, and they give information about the size of the highest peaks of the potential. The second relevant criterion is the probabilistic cost to confine the motion to the optimal area. The balance between the two strategies is subtle and will be described in detail in Sect. 3.2.

2.3.2 Moment Asymptotics Versus Almost Sure Asymptotics

Let us now explain the difference in the thinking about the annealed and the quenched setting. The asymptotics of the moments of $U(t)$ and its almost sure asymptotics are based on quite different (but related) arguments. The phenomenological difference between the two is the following.

First we consider the moments, i.e., the annealed setting, see also Remark 2.5. From the Feynman-Kac formula in (2.2) we see that the moments of $U(t)$ are the joint expectations over the path and over the potential. Hence, both random objects can 'work together' according to a joint strategy that is a compromise between the two; each of them gives a contribution that is exponentially costly: the potential assumes high values in a suitable area, and the path does not leave that area during the time interval $[0,t]$. For making the latter not too costly, the area should be a centred ball. Hence, the main contribution to the moments of $U(t)$ should come from a self-attractive behaviour of the random walk and an extreme behaviour of the potential. Phrasing it in terms of intermittency, it will turn out that the moments of the total sum over $z \in \mathbb{Z}^d$ of the solution $u(t,z)$ is asymptotically already well approximated by just the sub- sum on a much smaller region, which is centred at the origin and has a radius that we will call $R\alpha(t)$ in Chap. 3. This is an intermittent island, and we want to stress here that

> In the annealed setting, just one intermittent island is sufficient, and this island is centred.

Remark 2.11 (Estimating the probabilistic costs) Here is a simple rule of thumb for estimating the probabilistic cost for the random walk to stay in a ball of (t-dependent) radius $1 \ll r_t \ll \sqrt{t}$ until time t, the *non-exit probability*. Namely, it is of order $\mathrm{e}^{-O(t/r_t^2)}$. More precisely, writing '$\asymp$' if the quotient is bounded and bounded away from zero, we have

$$-\log \mathbb{P}_0(X([0,t]) \subset [-r_t, r_t]^d) \asymp \frac{t}{r_t^2}, \qquad t \to \infty, \tag{2.30}$$

where $\asymp$ means that the quotient of the two sides is bounded and bounded away from zero. This can be seen, with the help of the central limit theorem, as follows. Chop the random walk path into t/r_t^2 pieces of length r_t^2 to see that staying t time units in a ball with radius r is equivalent to each of these t/r_t^2 pieces staying in that ball. For each piece, the probability for doing this converges towards some fixed number in $(0,1)$, according to the central limit theorem. According to the Markov property, the total probability for all the pieces to stay within that ball, should be roughly the product of the single probabilities, i.e., a product of $O(t/r_t^2)$ terms of finite size in $(0,1)$, i.e, of size $\exp\{-O(t/r_t^2)\}$.

Another way to see that (2.30) holds (even with explicit identification of the prefactor) is to write $\mathbb{P}_0(X([0,t]) \subset [-r,r]^d)$ as the total mass $U_{B_r}(t)$ of the solution to the PAM with potential $\xi \equiv 0$ in $B_r = [-r,r]^d \cap \mathbb{Z}^d$ and to use the estimate in (2.21) to obtain that the left-hand side of (2.30) is not smaller than $-t\lambda_1(B_r) - \log(2r+1)^{d/2}$. Now one needs a rescaling argument for the principal eigenvalue $\lambda_1(B_r)$ of Δ^{d} in B_r; actually, one can show that $\lambda_1(B_r) \sim r^{-2}\lambda_1^{\mathrm{c}}([-1,1]^d)$ as $r \to \infty$, where $\lambda_1^{\mathrm{c}}([-1,1]^d)$ is the principal eigenvalue of the Laplace operator Δ in $[-1,1]^d$ with Dirichlet boundary condition. This shows that the left-hand side of (2.30) is even $\sim \lambda_1^c([-1,1]^d)\frac{t}{r_t^2}$. (Because of the appearance of the term $-\log(2r+1)^{d/2}$ in the above estimate, the upper limitation $r_t \ll \sqrt{t}$ has to be sharpened by adding a suitable logarithmic correction.)

Much more precise assertions are possible using the large-deviations principle in Lemma 4.3. ◇

In contrast, in the almost sure setting (see Chap. 5 for details), the quenched setting, the potential makes no particular effort of any kind, but behaves 'as usual'. The random path has to cope with that and must 'make the best' out of it. Hence, the identification of the almost sure asymptotics depends on a closer analysis of the potential landscape, almost surely for every sufficiently large t. In fact, within some large 'macrobox', one derives the existence of 'microboxes' (local, much smaller regions) in which $\Delta^{\mathrm{d}}+\xi$ possesses particularly large principal Dirichlet eigenvalues. Then one either lower bounds the Feynman-Kac formula for $U(t)$ by requesting the path to spend almost all its time there (neglecting the travel time to that place), or one lower bounds the principal eigenvalue of the macrobox against the local principal eigenvalue of one of these microboxes.

For the argument to work, one needs a good control on the upper tails of the eigenvalues of $\Delta^{\mathrm{d}}+\xi$ in local subregions inside a given large a priori box. In particular, one needs control on the probability that the local principal eigenvalues are extremely large. This is achieved by a control on their exponential moments with large prefactor, by use of the exponential Chebyshev inequality. This control in turn is a by-product of the proof of the asymptotics of the moments, since $\langle U(t)\rangle \approx \langle \mathrm{e}^{t\lambda_1}\rangle$. In this way, the analysis of the moments gives the necessary control on the upper tails of the eigenvalue and serves as an important input in the proof of the almost sure asymptotics.

In this way, we will use just one island (microbox) for a lower bound for the total mass. This estimate turns out to be very satisfactory, as it matches with a corresponding upper bound. However, this a priori does not mean that this island alone approximates the total mass $U(t)$ so well that the sum coming from the complement of this island is negligible with respect to the sum from that island. In order to achieve this, we a priori need to collect much more such islands, whose family is then called the *intermittent islands* in the sense of Sect. 1.4. Such assertions are handled with the help of spatial extreme-value analysis, which is used to identify the number, location, size and form of such islands. At this point, we would like to stress that

> In the quenched setting, there are a priori many intermittent islands, and they are widely spread, randomly located and much smaller than the annealed intermittent island.

It is another, much deeper story to prove that finally indeed just one of these islands (carefully picked) is sufficient to asymptotically exhaust the total mass, see Sect. 6.4.

The above explains only lower bounds (but very good ones). Most of the proofs in the literature for the corresponding upper bound do not reflect any details about the potential landscape and are quite abstract. We present the most successful proof strategies in Chap. 4.

2.3.3 *Mass Concentration*

In Chap. 6, we will go much deeper into the description of the PAM and will reveal much more information about the intermittent islands. This is well explained in terms of the eigenvalue expansion in (2.18):

$$u(t,\cdot) \approx u_{B^{(t)}}(t,\cdot) = \sum_{k=1}^{|B^{(t)}|} \mathrm{e}^{t\lambda_k} v_k(0) v_k(\cdot), \tag{2.31}$$

where $B^{(t)}$ is a centred box that is so large that the first approximation is good enough; see Remark 3.1.

The intermittent islands are equal to the regions where one of the eigenfunctions v_k has its main mass. More precisely, as Anderson localisation theory predicts, all the leading eigenfunctions v_k are highly concentrated in a region of small size somewhere in $B^{(t)}$ and are extremely close to zero everywhere outside. Each of these regions gives rise to a lower bound of the kind that we explained in Sect. 2.3.2. Furthermore, using extreme-value statistics, one can understand and prove that the shape of the potential ξ and the one of the solution $u(t,\cdot)$ in these islands approaches a certain deterministic form.

For the lower bound for $U(t)$ we just considered the first term in the above sum, the one with the largest eigenvalue. However, also the distance of the island to the starting point of the motion plays a rôle. Looking at the eigenvalue expansion in (2.31), this distance is roughly expressed by the term $v_k(0)$, by the fact that v_k is exponentially decreasing away from the centre of that island. Hence, the above heuristics gives the best lower bound by taking that k that maximises the term $\mathrm{e}^{t\lambda_k} v_k(0)$. This maximal k may be different from one. Then the conjecture is tempting that it is just this single summand that gives the overwhelming contribution, which is a rather strong form of intermittency, it is indeed a concentration assertion. This is indeed known for a number of potential distributions, see Chap. 6.

2.3.4 *Time-Evolution of the Mass Flow*

All heuristics so far considered only the situation of the mass flow at a given fixed, large time, i.e., a snapshot. However, one of the main goals is to describe the evolution of the mass flow, i.e., the function $t \mapsto u(t,\cdot)$. Making qualitative statements in this general view is rather difficult. However, in those cases in which the concentration property in just one island holds, we can make much more precise assertions. Indeed, we here can restrict the description of the main mass flow to a description of the time-evolution of the centre of that island. For this process there are explicit formulas available for a number of potential distributions; see Sect. 6.5.

An intuitive description of the main mass flow is as follows (see also [Mör11]). The box $B^{(t)}$ in (2.31), if it is picked sufficiently large, is the space horizon of the main mass at time t, i.e., the space in which it is located with high probability at time t. Now imagine that we look at a movie and let t increase, assuming that the radius of $B^{(t)}$ increases accordingly. Then from time to time it happens that this increasing horizon suddenly includes a new, much better local island than all islands that it contained before. Here 'better' refers to the relation between size of the local eigenvalue and the distance to the origin, as is expressed by the term $\mathrm{e}^{t\lambda_k} v_k(0)$. During a small time interval, this island becomes relevant and replaces the island that was optimal before. As a result, the main mass 'jumps' to the new island, and the Feynman-Kac formula is from now mainly concentrated on paths that go in short time to this new island and spend there most of the time.

There are two interesting time scales in this picture: the time lag during which a certain island is the optimal one, and the time lag during which the main mass moves from one optimal island to the next one. In all cases that we present in Sect. 6.5, the latter is much shorter than the first one. Furthermore, we also see there that the time lag during which an island is optimal increases from one optimal island to the next one, i.e., the mass flow *ages*; it behaves differently at late times than at early times. The effect of ageing will be highlighted and deepened in Sect. 6.5.

Chapter 3
Moment Asymptotics for the Total Mass

In this chapter, we explain what the asymptotics of the logarithm of the moments of the total mass $U(t)$ of the solution $u(t,\cdot)$ of the PAM in (1.1)–(1.2) are determined by, and how they can be described. This is fundamental for a deeper study of the PAM, and we will develop a rich picture. We will be working under the basic assumption that $(\xi(z))_{z\in\mathbb{Z}^d}$ is an i.i.d. random potential and that all positive exponential moments of $\xi(0)$ are finite, in which case all the moments of $U(t)$ are finite.

After making some basic observations in Sect. 3.1, we give in Sect. 3.2 a heuristic derivation, based on a large-deviation statement for the rescaled potential, under a crucial regularity assumption on the upper tails of $\xi(0)$ called *Assumption (J)*. This is followed by a second derivation in Sect. 3.3 in terms of a large-deviation statement for the local times of the random walk, under an assumption on the large-t behaviour of the logarithmic moment generating function $H(t)$ called *Assumption (H)*. We formulate the outcome of these heuristics for the respective potential distributions in Sect. 3.4. It turns out there that we need to distinguish four different regimes only, and we will provide explicit formulas in these regimes. The spatially continuous case is discussed in Sect. 3.5.

3.1 Rough Bounds

We recall the cumulant generating function $H(t) = \log\langle e^{t\xi(0)}\rangle$ of $\xi(0)$ defined in (1.13), which will play an important rôle here, since its behaviour as $t\to\infty$ describes the potential close to its essential supremum. Without much efforts, we obtain the two bounds

$$e^{H(t)-2dt} \le \langle U(t)\rangle \le e^{H(t)}, \qquad t\in(0,\infty). \tag{3.1}$$

W. König, *The Parabolic Anderson Model*, Pathways in Mathematics,
DOI 10.1007/978-3-319-33596-4_3

This easily follows from the Feynman-Kac formula $U(t) = \mathbb{E}_0[\mathrm{e}^{\int_0^t \xi(X_s)\,\mathrm{d}s}]$ in (2.3). Indeed, we obtain a first lower estimate for $\langle U(t)\rangle$ by restricting the expectation with respect to the random walk to the event $\bigcap_{s\in[0,t]}\{X_s = 0\}$ that it does not leave the origin up to time t. This event has probability e^{-2dt} as the time of the first jump, $\tau = \inf\{t > 0\colon X(t) \neq X(0)\}$, is exponentially distributed with parameter $2d$. Furthermore, on this event, we have that $\int_0^t \xi(X(s))\,\mathrm{d}s = t\xi(0)$. Hence,

$$\langle U(t)\rangle \geq \langle \mathbb{E}_0[\mathrm{e}^{t\xi(0)}\mathbb{1}_{\{\tau>t\}}]\rangle = \langle \mathrm{e}^{t\xi(0)}\rangle \mathrm{e}^{-2dt} = \mathrm{e}^{H(t)-2dt},$$

which shows the left inequality in (3.1). On the other hand, an upper estimate arises by applying Jensen's inequality in the exponential term in the Feynman-Kac representation to the probability measure on $[0,t]$ with Lebesgue density $1/t$ as follows:

$$\exp\Big\{\int_0^t \xi(X_s)\,\mathrm{d}s\Big\} \leq \int_0^t \frac{1}{t}\exp\Big\{t\xi(X_s)\Big\}\,\mathrm{d}s.$$

Now taking the expectation with respect to ξ and interchanging it with the integral over $\mathrm{d}s$ and the random walk expectation, we arrive at

$$\begin{aligned}\langle U(t)\rangle &\leq \Big\langle \int_0^t \frac{1}{t}\mathbb{E}_0\Big[\exp\Big\{t\xi(X_s)\Big\}\Big]\,\mathrm{d}s\Big\rangle = \int_0^t \frac{1}{t}\mathbb{E}_0\big[\langle \mathrm{e}^{t\xi(X_s)}\rangle\big]\,\mathrm{d}s = \langle \mathrm{e}^{t\xi(0)}\rangle \\ &= \mathrm{e}^{H(t)},\end{aligned} \tag{3.2}$$

which shows the right inequality in (3.1).

Because of (3.1), it appears appropriate to consider the term $\mathrm{e}^{-H(t)}\langle U(t)\rangle$ and to try to derive logarithmic asymptotics on the scale t or some smaller scale. This means that the moment asymptotics are described by (at least) two terms, the first of which is the cumulant generating function. This term yields a rough information about the way in which the potential attains large values, but no information about the structure of the potential in the high peaks. Therefore, we will have to work harder on the second term. Actually, it will turn out that it is more appropriate to replace $\mathrm{e}^{-H(t)}$ by some modification.

3.2 Heuristics via Eigenvalues

We give a heuristic derivation of a lower bound for $\langle U(t)\rangle$, which will later turn out to be also equal to the upper bound, up to the precision given by logarithmic asymptotics that we consider. However, the explanation of the lower bound here is intuitive and gives quite some insight in the behaviour of the PAM, while the proof of the upper bound does not. The main result of this section is (3.22).

Step 1: Catching in a large box. The first observation is that

$$\langle U(t)\rangle \sim \langle U_{B^{(t)}}(t)\rangle, \tag{3.3}$$

if the centred box $B^{(t)}$ is large enough, where we recall from Remark 1.6 that U_B denotes the total mass of the solution of the PAM in the set B. More details about (3.3) and the size of $B^{(t)}$ are given in Remark 3.1; for the remainder of this section it will be enough to know that the diameter of $B^{(t)}$ is large, but not larger than a power of t.

Remark 3.1 (Approximating with a large box) In order to approximate U with $U_{B^{(t)}}$ to obtain (3.3), one uses the Feynman-Kac formula (2.2) to see that

$$U(t) - U_B(t) = \mathbb{E}_0\Big[\mathrm{e}^{\int_0^t \xi(X_s)\,\mathrm{d}s}\mathbb{1}\{X_{[0,t]} \not\subset B\}\Big]. \tag{3.4}$$

This error term is small, if the box B is large, since it costs the path then much to travel to the outer boundary of B by time t. A qualitative upper bound is (see [GärMol98, Lemma 2.5(a)])

$$\mathbb{P}_0(X_{[0,t]} \not\subset [-R,R]^d) \leq 2^{d+1}\exp\Big\{-R\log\frac{R}{dt} + R\Big\}, \qquad R,t>0. \tag{3.5}$$

This estimate is particularly useful if R is much larger than t, as one gets a super-exponentially decaying upper bound. This can now be used in (3.4) in various ways, after taking the expectation w.r.t ξ (but also without). The easiest is to separate the two terms from each other by use of Hölder's inequality and afterwards using (3.5) for the probability term and (3.1) for the expectation. In this way, one arrives at some term in the exponential of the form $\frac{1}{p}H(pt)$ minus some term that comes from the right-hand side of (3.5). Depending on the asymptotics of $\frac{1}{p}H(pt) - H(t)$, one can pick R large enough that one can conclude that (3.3) holds. For many potentials including the double-exponential distribution, it suffices to take $R = R_t$ as $t(\log t)^{1+\eta}$ with some $\eta > 0$. ◇

Step 2: Switching to the principal eigenvalue. The second main step is the approximation

$$U_{B^{(t)}}(t) \approx \mathrm{e}^{t\lambda^{\mathrm{d}}(B^{(t)},\xi)} \tag{3.6}$$

(in the sense of logarithmic equivalence, i.e., the quotient of the logarithms converges to one), where we now write $\lambda^{\mathrm{d}}(B,\varphi)$ to denote the principal (i.e., largest) eigenvalue of the operator $\Delta^{\mathrm{d}} + \varphi$ in a finite set $B \subset \mathbb{Z}^d$ with zero boundary condition, for some potential $\varphi\colon B \to \mathbb{R}$ (hence $\lambda^{\mathrm{d}}(B,\xi) = \lambda_1(B)$ in the notation of Sect. 2.2.1). This approximation is a consequence of the spectral representation (2.18) and can be easily justified with the help of the methods that we described in Sect. 2.2.2. Hence, we have to understand the logarithmic asymptotics of the t-th exponential moments of the principal eigenvalue in a large, t-dependent box.

Remark 3.2 (p-th moments) It is already heuristically clear from (3.3) (and true in all known cases) that the p-th moments of $U(t)$ should have the same asymptotics as the first moments of $U(pt)$, at least as it concerns the leading terms. ◇

Remark 3.3 (Rough bounds on $\lambda^{\rm d}(B^{(t)},\xi)$) Observe that the leading eigenvalue $\lambda^{\rm d}(B^{(t)},\xi)$ is of the same order as the highest peak of ξ in $B^{(t)}$. Indeed, we easily check by the Rayleigh-Ritz-formula (2.20) that

$$\max_B \xi - 4d \le \lambda^{\rm d}(B,\xi) \le \max_B \xi, \qquad B \subset \mathbb{Z}^d \text{ finite.} \tag{3.7}$$

◇

Step 3: Adapting the potential to the intermittent island. We are going to introduce some decisive scales and to explain their interdependences. As we indicated in Sect. 2.2.3, the main contribution to $\langle {\rm e}^{t\lambda^{\rm d}(B^{(t)},\xi)}\rangle$ comes from realizations of the potential ξ having high peaks of some order $L(t)$ on mutually distant islands, the *intermittent islands*, whose radii are of some order $\alpha(t)$, which is much smaller than $\sqrt{t}$. As we now consider the expectation over the potential ξ, it will be much less costly on the probabilistic side to form just one such island and to place it around the origin. Furthermore, the potential will achieve the value $L(t)$ not precisely, but will have some deviation from that, whose order we will denote by $\gamma(t)$. More explicitly, we will consider the shifted and rescaled version of the potential,

$$\overline{\xi}_t(\cdot) = \gamma(t)\Big[\xi\big(\lfloor \cdot\,\alpha(t)\rfloor\big) - L(t)\Big]. \tag{3.8}$$

Scales for the moment asymptotics:

$\alpha(t)$ = order of diameter of intermittent island

$L(t)$ = maximal height of potential in the island

$\gamma(t)$ = reciprocal of the order of deviations of potential from $L(t)$ in the island

The appropriate orders of $\alpha(t)$, $L(t)$ and $\gamma(t)$ will be identified in (3.17), (3.18) and (3.19), respectively.

Remark 3.4 ($\alpha(t) \to \infty$) For definiteness, we are considering in this heuristic only the case where $\alpha(t) \to \infty$ as $t \to \infty$, which will be determined by Assumption (J) below. In the terminology of Sect. 3.4, we are here in one of the cases (AB) and (B). Hence, we need to rescale the intermittent island to $\alpha(t)$ times some 'continuous' subset of $\mathbb{R}^d$. By all means, there are interesting potential distributions that imply that $\alpha(t)$ does not diverge. This will turn out to be the cases (DE), where we put $\alpha(t) = 1$ and have islands of bounded size, and the case (SP), where $\alpha(t) \to 0$, since the island is a singleton. All the following formulas and reasonings have analogues. The decisive difference between cases (AB) and (B) on one side and (DE) and (SP) on the other is the need of a spatial rescaling in the two former cases. ◇

Here, in the case where $\alpha(t) \to \infty$, the core of our approach is the ansatz that $\overline{\xi}_t$ resembles some continuous shape function φ on the continuous box $Q_R = [-R, R]^d$ and that the outside of that box can be neglected. In other words, we use the lower bound

$$\begin{aligned} \langle \mathrm{e}^{t\lambda^{\mathrm{d}}(B^{(t)},\xi)} \rangle &\geq \Big\langle \mathrm{e}^{t\lambda^{\mathrm{d}}(B^{(t)},\xi)} \mathbb{1}\{\overline{\xi}_t \approx \varphi \text{ in } Q_R\} \Big\rangle \\ &\geq \Big\langle \mathrm{e}^{t\lambda^{\mathrm{d}}(B_{R\alpha(t)},\xi)} \mathbb{1}\{\overline{\xi}_t \approx \varphi \text{ in } Q_R\} \Big\rangle, \end{aligned} \tag{3.9}$$

where we write $B_R = [-R, R]^d \cap \mathbb{Z}^d$ for the discrete box with diameter $\approx 2R$. The second estimate follows from the monotonicity of the map $B \mapsto \lambda^{\mathrm{d}}(B, \xi)$. In (3.9), the potential ξ undertakes particular efforts in the 'microbox' $B_{R\alpha(t)}$ (equivalently, $\overline{\xi}_t$ in Q_R), and these will turn out to give the main contribution to the eigenvalue in the macrobox $B^{(t)}$, for proper choices for $\alpha(t)$, $L(t)$, $\gamma(t)$ and φ. This effort consists of assuming a particular shape φ inside this box, after proper rescaling. In the end we have to optimize over the box diameter R and over the potential shape φ.

Remark 3.5 (Large-deviation ansatz) Certainly, in (3.9) we are presenting an ansatz that turns out to lead to an optimal lower bound, in the sense that it matches with an upper bound, up to the precision that we consider. This is crucial and is typical for a large-deviation approach. It is based on a variant of the Laplace method (see Remark 7.1), which says that, among a sum of many exponential terms, the term with the largest rate wins. Here we use this idea in a very abstract manner (the 'sum' is indeed an integral over an enormously large space, the space of all potential shapes), and it is also technically rather cumbersome to make it work mathematically, but it makes the heuristics very instructive. In a more elaborate wording, we apply a large-deviation principle (LDP) for $\overline{\xi}_t$, see also (3.20) below. The theory of large deviations is instrumental to the study of the large-t asymptotics of the moments of $U(t)$, as we see here. See [DemZei98] for an account on the theory; in Sect. 4.2 we summarise the most important facts. ◇

Step 4: Identification of the scales. Let us find out what proper choices for $\alpha(t)$, $L(t)$ and $\gamma(t)$ are. For this purpose, we calculate the contribution from the event $\{\overline{\xi}_t \approx \varphi \text{ in } Q_R\}$. We have obviously

$$\overline{\xi}_t \approx \varphi \quad \text{in } Q_R \qquad \Longleftrightarrow \qquad \xi(\cdot) \approx L(t) + \tfrac{1}{\gamma(t)}\varphi\big(\tfrac{\cdot}{\alpha(t)}\big) \quad \text{in } B_{R\alpha(t)}. \tag{3.10}$$

It is clear that a shift of the potential by a constant shifts the eigenvalue by the same constant (and leaves the eigenfunction unchanged). Furthermore, it turns out that the only reasonable choice of $\gamma(t)$ is $\alpha(t)^2$, since the asymptotic scaling properties of the discrete Laplacian, $\Delta^{\rm d}$, imply that

$$\lambda^{\rm d}\big(B_{R\alpha(t)}, \tfrac{1}{\alpha(t)^2}\varphi\big(\tfrac{\cdot}{\alpha(t)}\big)\big) \approx \frac{1}{\alpha(t)^2}\lambda^{\rm c}(Q_R, \varphi), \tag{3.11}$$

where $\lambda^{\rm c}(Q, \varphi)$ denotes the principal eigenvalue of $\Delta + \varphi$ in a bounded set $Q \subset \mathbb{R}^d$ having a 'nice' boundary with zero boundary condition, and Δ is the usual 'continuous' Laplacian.

Remark 3.6 (Eigenvalue rescaling) The relation (3.11) even holds with '$\approx$' replaced by '$\sim$' and reflects the convergence of the discrete Laplace operator towards the continuous one after a spatial rescaling that is in the spirit of the central limit theorem. Its plausibility can be seen from the Rayleigh-Ritz principle in (2.20) by picking the (candidate for the) eigenfunction of the form $v(z) = \frac{1}{\alpha(t)^2}f\big(\frac{z}{\alpha(t)}\big)$ for some smooth function $f\colon \mathbb{R}^d \to [0, \infty)$ with support in Q_R. (Certainly, we have replaced B by $B_{R\alpha(t)}$ and ξ by $\frac{1}{\alpha(t)^2}\varphi(\frac{\cdot}{\alpha(t)})$.) Then '$\geq$' in (3.11) is more or less clear by optimising over v. However, proving the opposite inequality needs some work that can be carried out using techniques from the theory of Gamma-convergence, a theory from variational analysis that deals with the interchanging of limits and suprema. See [Bra02] for a rather readable introduction to this theory. ◇

Because of (3.11), we need to choose $\gamma(t) = \alpha(t)^2$ and obtain, on the event $\{\overline{\xi}_t \approx \varphi \text{ in } Q_R\}$,

$$\mathrm{e}^{t\lambda^{\rm d}(B_{R\alpha(t)}, \xi)} \approx \mathrm{e}^{tL(t)} \exp\Big\{\frac{t}{\alpha(t)^2}\lambda^{\rm c}(Q_R, \varphi)\Big\}. \tag{3.12}$$

That is, we extracted from the eigenvalue term a leading scale (just $L(t)$) and a second scale, $t\alpha(t)^{-2}$, with an interesting prefactor that we will have to optimise on. Now we need to choose $\alpha(t)$ and $L(t)$ such that the probability of the event $\{\overline{\xi}_t \approx \varphi \text{ in } Q_R\}$ is also on the exponential scale $t\alpha(t)^{-2}$. (Otherwise, one of the two scales will be negligible w.r.t. the other, and after optimisation we will obtain a trivial prefactor, 0 or ∞.) Hence, we have to make a suitable assumption on the tails of the potential distribution.

Assumption (J) *There is a positive scale function η and a strictly increasing function J such that*

$$\lim_{s\to\infty}\frac{1}{\eta(s)}\log\mathrm{Prob}\Big(\xi(0)>\frac{H(s)}{s}+\frac{\eta(s)}{s}x\Big)=-J(x),\qquad x\in\mathbb{R}. \tag{3.13}$$

On the left-hand side, we may replace '>' by '≈' (with some meaning that we do not want to specify here), by strict monotonicity of J. Under Assumption (J), we calculate

$$\begin{aligned}\mathrm{Prob}\Big(\xi(\cdot)\approx\tfrac{H(s)}{s}+\tfrac{\eta(s)}{s}\varphi\big(\tfrac{\cdot}{\alpha(t)}\big)\text{ in }B_{R\alpha(t)}\Big)&\approx\prod_{z\in B_{R\alpha(t)}}\mathrm{e}^{-\eta(s)J\left(\varphi(\frac{z}{\alpha(t)})\right)}\\&\approx\exp\Big\{-\eta(s)\alpha(t)^dI_R(\varphi)\Big\},\end{aligned} \tag{3.14}$$

where

$$I_R(\varphi)=\int_{Q_R}J(\varphi(y))\,\mathrm{d}y, \tag{3.15}$$

using that the potential ξ is i.i.d., and after turning the Riemann sum of the $J(\varphi(\cdot))$-values into an integral. In order that this potential shape scaling coincides with our ansatz in (3.11) and that the scale of this probability coincides with the one in (3.12), we need to introduce a new scale function $s(t)$ such that

$$\frac{\eta(s(t))}{s(t)}=\frac{1}{\alpha(t)^2}\qquad\text{and}\qquad\frac{t}{\alpha(t)^2}=\eta(s(t))\alpha(t)^d. \tag{3.16}$$

Clearly, this requires that $s(t)=t\alpha(t)^{-d}$. Hence, $\alpha(t)$ is determined by the requirement

$$\frac{\eta\big(t\alpha(t)^{-d}\big)}{t\alpha(t)^{-d}}=\frac{1}{\alpha(t)^2}, \tag{3.17}$$

which in turn implies that

$$L(t)=\frac{H(t\alpha(t)^{-d})}{t\alpha(t)^{-d}}. \tag{3.18}$$

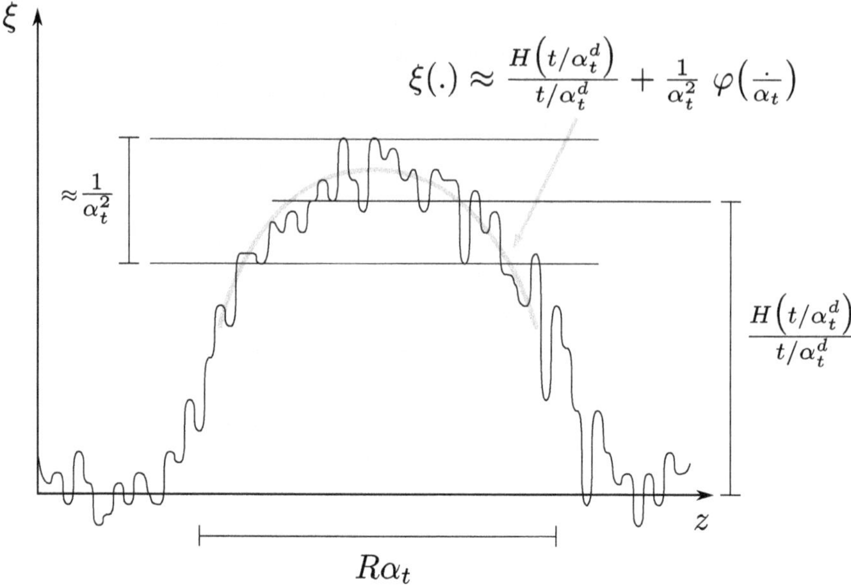

Fig. 3.1 The event $\{\overline{\xi}_t \approx \varphi \text{ in } Q_R\}$: the random potential ξ approaches a certain shifted, rescaled version of the profile φ (*shaded line*) in the box with radius $R\alpha(t)$

Step 5: Finish. Hence, we have identified the right scales and see that

$$\overline{\xi}_t(\cdot) = \alpha(t)^2\Big(\xi(\lfloor\cdot\,\alpha_t\rfloor) - \frac{H(t\alpha(t)^{-d})}{t\alpha(t)^{-d}}\Big) \tag{3.19}$$

formally satisfies a *large-deviation principle (LDP)* with speed $t\alpha(t)^{-2}$ and rate function I_R defined in (3.15) in the box Q_R, i.e., in simple terms,

$$\operatorname{Prob}(\overline{\xi}_t \approx \varphi \text{ in } Q_R) \approx \exp\Big\{-\frac{t}{\alpha(t)^2} I_R(\varphi)\Big\}. \tag{3.20}$$

The event $\{\overline{\xi}_t \approx \varphi \text{ in } Q_R\}$ is depicted in Fig. 3.1; recall (3.10) and (3.19).

So we arrive at

$$\begin{aligned}\big\langle U(t)\big\rangle &\geq \Big\langle \mathrm{e}^{t\lambda^{\mathrm{d}}(B_{R\alpha(t)},\xi)}\, \mathbb{1}\{\overline{\xi}(\cdot) \approx \varphi(\cdot) \text{ in } Q_R\}\Big\rangle \\ &\approx \mathrm{e}^{\alpha(t)^d H(t/\alpha(t)^d)} \exp\Big\{\frac{t}{\alpha(t)^2}\lambda^{\mathrm{c}}(Q_R,\varphi)\Big\}\operatorname{Prob}(\overline{\xi}_t \approx \varphi \text{ in } Q_R) \\ &\approx \mathrm{e}^{\alpha(t)^d H(t/\alpha(t)^d)} \exp\Big\{-\frac{t}{\alpha(t)^2}\Big(I_R(\varphi) - \lambda^{\mathrm{c}}(Q_R,\varphi)\Big)\Big\}.\end{aligned} \tag{3.21}$$

Optimising on φ and R, we finally obtain the main result of these heuristics:

Under Assumption (J),

$$\big\langle U(t)\big\rangle \geq \mathrm{e}^{\alpha(t)^d H(t/\alpha(t)^d)} \exp\Big\{-\frac{t}{\alpha(t)^2}(\chi+o(1))\Big\}, \qquad t\to\infty, \tag{3.22}$$

where the constant χ is given in terms of the characteristic variational problem

$$\begin{aligned}\chi &= \lim_{R\to\infty} \inf_{\varphi\in\mathcal{C}(Q_R)} \big[I_R(\varphi) - \lambda^{\mathrm{c}}(Q_R,\varphi)\big] \\ &= \inf_{\varphi\in\mathcal{C}(\mathbb{R}^d)} \Big[\int_{\mathbb{R}^d} J\circ\varphi - \sup_{g\in H^1(\mathbb{R}^d):\|g\|_2=1} \Big(\int_{\mathbb{R}^d} \varphi g^2 - \|\nabla g\|_2^2\Big)\Big].\end{aligned} \tag{3.23}$$

Here we used the Rayleigh-Ritz principle (2.20) for the principal eigenvalue, and $\mathcal{C}(Q)$ is the set of all continuous functions $Q\to\mathbb{R}$.

We arrived at the intended lower bound for the main assertion on the moment asymptotics. Recall that we did this heuristics under the assumption that $\alpha(t)\to\infty$; in the other cases the formula (3.23) must be replaced by a discrete version. In Sect. 3.4 we summarize the moment asymptotics and give the characteristic formulas in all these cases.

The upper bound '$\leq$' also holds in (3.22), but its proof is much more technical and difficult and does not give insight in the behaviour of the model. See Sect. 4 for some techniques for proving the upper bound.

Remark 3.7 (Interpretation of χ) The variational formula on the right of (3.22) is an important object for the description of the long-time behaviour of the PAM, as it contains interesting information about the behaviour of the potential in the intermittent island. The first term is determined by the absolute height of the typical realizations of the potential, and the second contains information about the shape of the potential close to its maximum in spectral terms of the Anderson Hamiltonian $\Delta^{\mathrm{d}}+\xi$ in this region. More precisely, those realizations of ξ with $\overline{\xi}_t\approx\varphi_*$ in Q_R for large R and φ_* a minimizer in the variational formula in (3.23) contribute most to $\langle U(t)\rangle$. In particular, the geometry of the relevant potential peaks is hidden via χ in the second asymptotic term of $\langle U(t)\rangle$. Hence, χ is a *characteristic formula* that gives, via an optimisation, a lot of useful information, which will be instrumental for a number of deeper investigations. ◇

Remark 3.8 (Potential confinement properties) The above heuristics suggest, in the spirit of large-deviation theory, that the main contribution to the moments should come from those realisations of the potential ξ such that the rescaled shifted version

$\overline{\xi}_t$ resembles the members of the set $\mathcal{M}$ of minimizer(s) of the variational formula in (3.23). This can be formulated in terms of a kind of law of large numbers for $\overline{\xi}_t$ under the *annealed potential measures*

$$\widehat{Q}_t(\mathrm{d}\varphi) = \frac{1}{\langle U(t)\rangle}\mathbb{E}_0\Big[\exp\Big\{\int_0^t \xi(X_s)\,\mathrm{d}s\Big\}\Big]\mathrm{Prob}(\xi \in \mathrm{d}\varphi), \qquad t \geq 0,\ \varphi\colon \mathbb{Z}^d \to \mathbb{R}; \tag{3.24}$$

note the analogy to the annealed path measure in (2.16). More precisely, if $U(\mathcal{M})$ is some neighbourhood of $\mathcal{M}$ in a suitable topology, then the event $\{\xi\colon \overline{\xi}_t \notin U(\mathcal{M})\}$ should be of asymptotically small probability w.r.t. $\widehat{Q}_t$. (We got around specifying that topology by writing '$\overline{\xi}_t \approx \varphi$' in the preceding.)

Such a property is called a *potential confinement property*. There is no doubt that such a law of large numbers should be valid in great generality, but there are only few formulations or even proofs for this in the literature. Such a statement has been proved in the case of an almost bounded potential ξ from the class (AB) (in the notation introduced in Sect. 3.4 below) in [GrüKön09], and there exists a quenched version of such statement for the double-exponential distribution, due to [GärKönMol07], which we report on in Sect. 6.2.

A technical problem for the proof is that the problem in (3.23), and hence also $\mathcal{M}$, is spatially shift-invariant, i.e., one has to cope with the event that $\overline{\xi}_t$ does not resemble any shift of the minimiser(s). Another problem is to prove a certain *stability* of the characteristic variational problem (3.23), the property that says that, for any sequence φ_n of admissible functions such that the functional in the second line converges to its infimum, there is a subsequence that converges towards the minimiser, up to spatial shifts. This property needs to be valid in the same topology in which the large deviation principles under Q_t hold. ◇

3.3 Heuristics via Local Times

In this section, we present another route along which the (lower bound of the) asymptotics of the moments of $U(t)$ can be identified. This route is in a sense 'dual' to the route that we described in Sect. 3.2: instead of carrying out the expectation with respect to the random walk in the Feynman-Kac formula in (2.2) first, and then analysing the ξ-expectation of the resulting expression in the eigenvalue expansion, we start by carrying out the ξ-expectation and then analyse the resulting expectation over the random walk. The main result of this section is in (3.31).

To that end, we recall from (2.11) one of the main objects in the probabilistic treatment of the PAM, the *local times* $\ell_t(z) = \int_0^t \delta_z(X_s)\,\mathrm{d}s$ of the random walk

$(X_s)_{s\in[0,\infty)}$. Again, the cumulant generating function $H(t) = \log\langle \mathrm{e}^{t\xi(0)}\rangle$ plays a major rôle; it is again assumed to be finite for any $t > 0$. From (2.13) we already know that

$$\langle U(t)\rangle = \mathbb{E}_0\Big[\exp\Big\{\sum_{z\in\mathbb{Z}^d} H(\ell_t(z))\Big\}\Big]. \tag{3.25}$$

We are going to work under the following supposition on the asymptotics of H.

Assumption (H) *There are a function* $\widehat{H}\colon(0,\infty) \to \mathbb{R}$ *and a continuous function* $\eta\colon(0,\infty)\to(0,\infty)$ *such that*

$$\lim_{t\uparrow\infty}\frac{H(ty)-yH(t)}{\eta(t)} = \widehat{H}(y) \neq 0 \qquad \textit{for } y\neq 1, \tag{3.26}$$

and the limit $\eta_* = \lim_{t\to\infty}\eta(t)/t \in [0,\infty]$ *exists.*

Assumption (H) is crucial and will be discussed at length in Sect. 3.4 below. It is essentially equivalent to Assumption (J), as we discuss in Remark 3.11. Let us already remark that the function η coincides with the one of Assumption (J) above.

We are certainly going to use the same scales as in Sect. 3.2, but we have to determine them now by means of Assumption (H). As above, we define the scale function $\alpha(t)$ by

$$\frac{\eta\big(t\alpha(t)^{-d}\big)}{t\alpha(t)^{-d}} = \frac{1}{\alpha(t)^2}. \tag{3.27}$$

As in Sect. 3.2, we restrict to the case $\alpha(t)\to\infty$. Then we need to consider the spatially rescaled version of the local times,

$$L_t(y) = \frac{\alpha(t)^d}{t}\ell_t(\lfloor\alpha_t y\rfloor), \qquad y\in Q_R = [-R,R]^d, \tag{3.28}$$

which is a random, L^1-normalised step function. In order to make use of (3.26), we now continue (3.25) with

$$\begin{aligned}\langle U(t)\rangle &= \mathrm{e}^{\alpha(t)^d H(t/\alpha(t)^d)}\\ &\quad\times\mathbb{E}_0\Big[\exp\Big\{\eta\big(\tfrac{t}{\alpha(t)^d}\big)\sum_{z\in\mathbb{Z}^d}\frac{H\big(L_t(\frac{z}{\alpha(t)})\frac{t}{\alpha(t)^d}\big)-L_t(\frac{z}{\alpha(t)})H(\frac{t}{\alpha(t)^d})}{\eta(\frac{t}{\alpha(t)^d})}\Big\}\Big].\end{aligned}$$

The validity of this is easy to verify, using that $\sum_{z\in\mathbb{Z}^d} \ell_t(z) = t$. By definition of α in (3.27), we may replace the prefactor in the exponent by $t/\alpha(t)^{d+2}$. According to Assumption (H), we may asymptotically replace the quotient after the sum on z by $\widehat{H}(L_t(z/\alpha(t)))$. Turning this sum into an integral using the substitution $z = \lfloor y\alpha(t)\rfloor$, we arrive at

$$\begin{aligned}\langle U(t)\rangle &\approx \mathrm{e}^{\alpha(t)^d H(t/\alpha(t)^d)}\mathbb{E}_0\Big[\exp\Big\{\frac{t}{\alpha(t)^2}\frac{1}{\alpha(t)^d}\sum_{z\in\mathbb{Z}^d}\widehat{H}(L_t(z/\alpha(t)))\Big\}\Big]\\ &\approx \mathrm{e}^{\alpha(t)^d H(t/\alpha(t)^d)}\mathbb{E}_0\Big[\exp\Big\{\frac{t}{\alpha(t)^2}\int_{\mathbb{R}^d}\widehat{H}(L_t(y))\,\mathrm{d}y\Big\}\Big].\end{aligned} \tag{3.29}$$

We see that, after all these elementary manipulations, we arrived at some interesting term that has an exponential rate on the scale $t\alpha(t)^{-2}$, which is precisely the scale that we encountered in Sect. 3.2, see (3.20). A crucial fact is that $(L_t)_{t\in(0,\infty)}$ satisfies a large-deviation principle (LDP) with speed $t\alpha(t)^{-2}$ and rate function $g^2 \mapsto \|\nabla g\|_2^2$, that is,

$$\mathbb{P}_0(L_t(\cdot) \approx g^2(\cdot) \text{ in } Q_R) \approx \exp\Big\{-\frac{t}{\alpha(t)^2}\|\nabla g\|_2^2\Big\}, \tag{3.30}$$

for any L^2-normalised function $g \in H^1(\mathbb{R}^d)$ with support in Q_R (see Sect. 4.2 for details). Now we obtain a lower bound in (3.29) by inserting the indicator on the event $\{L_t(\cdot) \approx g^2(\cdot) \text{ in } Q_R\}$ and optimising over g^2 and R. This implies the main result of these heuristics:

Under Assumption (H),

$$\langle U(t)\rangle \geq \mathrm{e}^{\alpha(t)^d H(t/\alpha(t)^d)}\exp\Big\{-\frac{t}{\alpha(t)^2}(\chi_\circ + o(1))\Big\}, \tag{3.31}$$

where $\chi_\circ$ is given as

$$\chi_\circ = \inf\Big\{\|\nabla g\|_2^2 - \int_{\mathbb{R}^d}\widehat{H}\circ g^2 \colon g \in H^1(\mathbb{R}^d),\ \|g\|_2 = 1\Big\}. \tag{3.32}$$

Observe that this is 'dual' to (3.22) in the sense that the two variational formulas χ and $\chi_\circ$ are 'dual' to each other, see Remark 3.12. The remarks below (3.23) apply here as well.

Remark 3.9 (Interpretation of $\chi_\circ$) The characteristic variational formula in (3.32) has an analogous interpretation and importance as the formula χ in (3.23). Here it is the (rescaled) local times L_t in the Feynman-Kac formula that is considered instead of the random potential. One main information that one should grasp is that the main contribution to the expectation in (3.25) comes from those paths whose rescaled local times L_t resemble the minimiser(s) g^2 in (3.32). This assertion is not innocent, as the set of minimisers, $\mathcal{M}_\circ$, is shift-invariant, i.e., any spatial shift of a minimiser is also a minimiser. Hence, even if one would know that, in some sense, L_t belongs to a neighbourhood of $\mathcal{M}_\circ$ with high probability (like for χ, one calls this a *tube property*), the question arises to which shift of the minimiser L_t is attracted. See Remark 3.14 for more about that. ◇

Remark 3.10 (Higher moments) Handling the p-th moment of the total mass $U(t)$ in the above way presents no problem, as long as p is an integer. Indeed, one then writes the Feynman-Kac formula as an expectation over p independent random walks and proceeds with a large-deviations principle for the vector of these p motions, see [GärMol98], e.g. For non-integer p, one has to employ convexity equations in a clever way, see [GärKön00, Lemma 1] for $p \geq 1$ and [BisKön01, Lemma 4.1] for $p \in (0,1)$. The outcome is that, at least up to the precision that we consider in (3.31), the asymptotics of $\langle U(t)^p\rangle$ are identical to the ones of $\langle U(tp)\rangle$, as we mentioned and motivated in Remark 3.2. ◇

Remark 3.11 ((J)$\overset{?}{\Longleftrightarrow}$(H)) Let us comment on the relation between the two Assumptions (J) and (H) under which we derived the two main assertions (3.22) and (3.31). First note that the scale functions η appearing in (3.13) and (3.26) are identical. Essentially, the crucial connection between Assumptions (J) and (H) is that they describe the second-order terms of the upper tails of $\xi(0)$ and of the exponential moments, respectively. Introducing the t-dependent random variable $X_t = (\xi(0) - H(t)/t)t/\eta(t)$, we see that (J) describes the upper tails of X_t:

$$(3.13) \qquad \Longleftrightarrow \qquad \lim_{t\to\infty} \frac{1}{\eta(t)} \log \operatorname{Prob}(X_t > x) = J(x), \quad x \in \mathbb{R}, \tag{3.33}$$

and, because of $\langle e^{\eta(t)yX_t}\rangle = \langle e^{ty\xi(0)}\rangle e^{-yH(t)} = e^{H(ty)-yH(t)}$, (H) describes the exponential moments of X_t:

$$(3.26) \qquad \Longleftrightarrow \qquad \lim_{t\to\infty} \frac{1}{\eta(t)} \log\langle e^{\eta(t)yX_t}\rangle = \widehat{H}(y), \quad y > 0. \tag{3.34}$$

We now claim that Assumption (H) implies Assumption (J), but the reversed implication is true only under the following additional assumption on the very large values of X_t:

$$\lim_{M\to\infty} \lim_{t\to\infty} \frac{1}{\eta(t)} \log \big\langle e^{\eta(t)yX_t} \mathbb{1}\{X_t > M\}\big\rangle = -\infty, \qquad y > 0. \tag{3.35}$$

Furthermore, $\widehat{H}$ is the Legendre transform of J, that is,

$$\widehat{H}(y) = \max_{x>0}(xy - J(x)), \qquad y > 0. \tag{3.36}$$

Let us sketch the proofs of these assertions; they are fundamental in the theory of large deviations and in other parts of mathematics (e.g. in the theory of regular functions). The derivation of Assumption (J) from (H) decomposes into proving the upper bound on the right of (3.33) and the lower bound. The upper bound is easily derived via an elementary application of the *exponential Chebychev inequality*, which is the *Markov inequality* for the function $x \mapsto e^x$. Indeed, for any $y > 0$ we have

$$\begin{aligned}\text{Prob}(X_t > x) &= \text{Prob}(e^{\eta(t)yX_t} > e^{yx\eta(t)}) \leq \langle e^{\eta(t)yX_t}\rangle e^{-yx\eta(t)} \approx e^{\eta(t)\widehat{H}(y)}e^{-yx\eta(t)} \\ &= e^{-\eta(t)[yx-\widehat{H}(y)]},\end{aligned}$$

where '$\approx$' denotes equality up to an error $e^{o(\eta(t))}$. Now the best upper bound that we can get in this way is to take the maximum over $y > 0$ over the expression within $[\cdots]$ in the exponent, and an upper bound with the Legendre transform of $\widehat{H}$ arises. But this is equal to J by duality of Legendre transforms.

The proof of the lower bound follows the well-known proof of the *Gärtner-Ellis theorem*, see [DemZei98]. The proof needs a transformed measure $\text{Prob}_{x,t}$ with density $e^{\eta(t)y_xX_t}\langle e^{\eta(t)y_xX_t}\rangle^{-1}$, where $y_x > 0$ is chosen as a maximiser in the Legendre transform relation $J(x) = \max_y[yx - \widehat{H}(y)]$. For technical reasons, we have to do that for $x + \varepsilon$ instead of x, for some auxiliary $\varepsilon > 0$. Then one rewrites

$$\begin{aligned}\text{Prob}(X_t > x) &= \langle e^{\eta(t)y_{x+\varepsilon}X_t}\rangle \Big\langle e^{-\eta(t)y_{x+\varepsilon}X_t}\mathbb{1}\{X_t > x\}\Big\rangle_{x+\varepsilon,t} \\ &\geq e^{\eta(t)(\widehat{H}(y_{x+\varepsilon})+o(1))}\Big\langle e^{-\eta(t)y_{x+\varepsilon}X_t}\mathbb{1}\{x + 2\varepsilon > X_t > x\}\Big\rangle_{x+\varepsilon,t} \\ &\geq e^{\eta(t)(\widehat{H}(y_{x+\varepsilon})+o(1))}e^{-\eta(t)y_{x+\varepsilon}(x+2\varepsilon)}\text{Prob}_{x+\varepsilon,t}(x + 2\varepsilon > X_t > x),\end{aligned}$$

where the second step used the right assertion of (3.34). Leaving the ε and the last probability term aside, we already arrived at the desired estimate, since $\widehat{H}(y_x) - y_xx = -J(x)$, according to the Legendre relation. The main problem is to show that the last probability term has the exponential rate zero on the scale $\eta(t)$. The best way to do that is to show that it converges towards one. This decomposes into two parts, the 'upper' of which we show now. Again applying the exponential Chebyshev inequality, we obtain, for any $r > 0$,

$$\begin{aligned}\mathrm{Prob}_{x+\varepsilon,t}(X_t \geq x + 2\varepsilon) &\leq \langle \mathrm{e}^{\eta(t)rX_t}\rangle_{x+\varepsilon,t}\mathrm{e}^{-r\eta(t)(x+2\varepsilon)} \\ &\approx \exp\Big\{-\eta(t)\Big[-\widehat{H}(y_{x+\varepsilon}+r)+\widehat{H}(y_{x+\varepsilon})-r(x+2\varepsilon)\Big]\Big\}.\end{aligned}$$

Now using that $\widehat{H}'(y_x) = x$ by maximality of y_x, and employing a Taylor expansion around y_x, one sees that, by proper choice of r, the term in $[\cdots]$ is positive. Hence, the probability of the event $\{X_t \geq x + 2\varepsilon\}$ vanishes. Similarly one shows the same about the event $\{X_t \leq x\}$, which finishes the proof.

The proof that, under (3.35), Assumption (H) follows from Assumption (J) is done by proving that Assumption (J) and the strict monotonicity of J imply that X_t satisfies a large-deviation principle with scale $\eta(t)$ and rate function J (this is done similarly to the respective part of the proof of Cramér's theorem), and then the right-hand assertion of (3.34) follows from Varadhan's lemma, using (3.35). See [DemZei98] for details. ◇

Remark 3.12 ($\chi_\circ = \chi$) In view of the fact that in both (3.21) and (3.31) also the complementary inequality $\leq$ holds, the two variational formulas in (3.23) and (3.32) must be identical. This can also be seen in an analytical way by interchanging the infimum and the supremum in (3.23):

$$\begin{aligned}\chi &= \inf_{g\in H^1(\mathbb{R}^d):\|g\|_2=1}\ \inf_{\varphi\in C(\mathbb{R}^d)}\Big[\int_{\mathbb{R}^d} J\circ\varphi - \Big(\int_{\mathbb{R}^d}\varphi g^2 - \|\nabla g\|_2^2\Big)\Big] \\ &= \inf_{g\in H^1(\mathbb{R}^d):\|g\|_2=1}\Big[\|\nabla g\|_2^2 - \sup_{\varphi\in C(\mathbb{R}^d)}\Big(\int_{\mathbb{R}^d}\varphi g^2 - \int_{\mathbb{R}^d} J\circ\varphi\Big)\Big] \\ &= \chi_\circ.\end{aligned} \tag{3.37}$$

To see that the last step holds, one has to show that

$$\int_{\mathbb{R}^d}\widehat{H}\circ g^2 = \sup_{\varphi\in C(\mathbb{R}^d)}\Big(\int_{\mathbb{R}^d}\varphi g^2 - \int_{\mathbb{R}^d} J\circ\varphi\Big),$$

i.e., the functional $g^2 \mapsto \int H\circ g^2$ is the Legendre transform of the map $\varphi \mapsto \int J\circ\varphi$. But this is easily derived from the relation (3.36). ◇

Hence, the heuristics of this and the preceding section lead us to the formulation of the following result. We also include the case where $\alpha(t)$ does not diverge as $t \to \infty$, which we excluded from the above heuristics.

Theorem 3.13 (Moment asymptotics) *Suppose that Assumption (H) holds, and define* $\alpha(t)$ *by* (3.17).

(1) Assume that $\lim_{t\to\infty}\eta(t)/t = 0$. *Then* $\alpha(t)\to\infty$, *and*

$$\langle U(t)\rangle = \mathrm{e}^{\alpha(t)^d H(t/\alpha(t)^d)} \exp\Big\{-\frac{t}{\alpha(t)^2}(\chi_\circ + o(1))\Big\}, \tag{3.38}$$

with $\chi_\circ = \chi$ *given by* (3.23) *and* (3.32).

(2) Assume that $\lim_{t\to\infty}\eta(t)/t = 1$. *Then* $\alpha(t)\to 1$, *and* $\widehat{H}(y) = \rho y\log y$ *for some* $\rho\in(0,\infty)$, *and*

$$\langle U(t)\rangle = \mathrm{e}^{H(t)} \exp\big\{-t(\chi_\rho + o(1))\big\}, \tag{3.39}$$

with χ_ρ *given by* (3.45) *and* (3.47), *the spatially discrete version of* (3.23) *and* (3.32), *respectively (see Remark 3.17).*

(3) Assume that $\lim_{t\to\infty}\eta(t)/t = \infty$. *Then* $\alpha(t)\to 0$ *and* (3.39) *holds with* $\rho=\infty$, *where* $\chi_\infty = 2d$ *is also given by* (3.45) *and* (3.47) *(see Remark 3.16).*

In case (1) the diameter $\alpha(t)$ of the intermittent island for the moments diverges, in case (2) it stays bounded and bounded away from zero, and in case (3) the island shrinks to a single site.

Theorem 3.13 has been proved in the literature for various choices of the distribution of the potential ξ, but the proofs are scattered over a number of papers. See Sect. 3.4 for references and explicit formulas. A rigorous proof under Assumption (J) does not seem to be given yet.

Remark 3.14 (Path confinement properties) Analogously to Remark 3.8, it is also tempting to guess that the rescaled local times should satisfy a law of large numbers, i.e., should converge to the set $\mathcal{M}_\circ$ of minimiser(s) in (3.32) in probability with respect to the transformed path measure given in (2.15). Explicitly: for any neighbourhood $U(\mathcal{M}_\circ)$ of $\mathcal{M}_\circ$,

$$\limsup_{t\to\infty}\frac{\alpha(t)^2}{t}\log\frac{\mathbb{E}_0\big[\exp\big\{\sum_{z\in\mathbb{Z}^d}H(\ell_t(z))\big\}\mathbb{1}\{L_t\in U(\mathcal{M}_\circ)^{\mathrm{c}}\}\big]}{\mathbb{E}_0\big[\exp\big\{\sum_{z\in\mathbb{Z}^d}H(\ell_t(z))\big\}\big]} < 0. \tag{3.40}$$

This property is called *path confinement property*. Sometimes also the name *tube property* is used, as the set $\mathcal{M}_\circ$ can be thought of as a line if it consists only of all the shifts of some unique function, and then a typical neighbourhood $U(\mathcal{M}_\circ)$ is a tube around that line.

The topology in this statement and the one of the neighbourhood $U(\mathcal{M}_\circ)$ must coincide with the one of the LDP in (3.30).

Proving the tube property in (3.40) requires two main steps (and a lot of technicalities). According to large-deviation theory (notably (3.29), the LDP in (3.30) and Varadhan's lemma, see Sect. 4.2), one expects that

$$\text{l.h.s. of (3.40)} = -\inf_{g^2\in U(\mathcal{M}_\circ)^c}\Big(\|g\|_2^2-\int \widehat{H}\circ g^2\Big)+\inf_{g^2}\Big(\|g\|_2^2-\int \widehat{H}\circ g^2\Big). \quad (3.41)$$

Proving (3.41) is involved because of the lack of compactness in the LDP, an ubiquitous problem that we explain in Sect. 4.2. One has to extend compactification techniques, notably the periodisation technique outlined in Sect. 4.3, or one has to adapt some of the techniques that we present in Chap. 4. This is the first major step on a way to a proof of (3.40).

The second step is to prove that the right-hand side is strictly negative. This is usually done with the help of stability, once this property has been established. (We recall that this is the property of a variational problem that its minimiser(s) can be approached only by minimising the value of the functional.) This is a serious piece of work that has to be done on a case-by-case basis. See Sect. 7.1 for examples of variational problems of the form of $\chi_\circ$, whose stability has been established and has been exploited for deriving deeper properties of the PAM.

For a related model, the above has been carried out in [BolSch97]. For the PAM with doubly-exponentially distributed potential, a version of (3.40) can be derived from results of [GärHol99], which we outline in Sect. 7.1.1. There we also go deeper and reveal the distribution of the shift to which the path is attracted. In that example, which lives in $\mathbb{Z}^d$ rather than in $\mathbb{R}^d$, $\mathcal{M}_\circ$ consists of just one function, modulo shifts, and L_t converges in distribution under the annealed path measure towards a random shift of that minimiser, whose distribution is identified explicitly. ◇

3.4 Decomposition into Four Potential Classes

In this section we explain that, under the crucial regularity Assumption (H), there are only four regimes (called universality classes in [HofKönMör06]) of asymptotic behaviours of the PAM. Turning it around, the regular potentials decompose into four classes, each of which leads to a significantly different asymptotic behaviour of the PAM. Each of these behaviours comes with a spatial scale ($\alpha(t)$) and an explicit characteristic variational formula, given in two forms that are dual to each other: one describing the shape of the potential (χ) and one describing the shape of the local times ($\chi_\circ$). We follow [HofKönMör06].

Recall Assumption (H) from Sect. 3.2, see (3.26). The function $\widehat{H}$ extracts the asymptotic scaling properties of the cumulant generating function H. In the language of the theory of regular functions, the assumption is that the logarithmic moment generating function H is in the *de Haan class* (see [BinGolTeu87] for more on the theory of regular functions). This does not leave many possibilities for $\widehat{H}$:

Proposition 3.15 *Suppose that Assumption (H) holds.*

(i) There is a $\gamma \geq 0$ *such that* $\lim_{t\uparrow\infty} \eta(yt)/\eta(t) = y^\gamma$ *for any* $y > 0$, *i.e.,* η *is regularly varying of index* γ. *In particular,* $\eta(t) = t^{\gamma+o(1)}$ *as* $t \to \infty$.

(ii) There exists a parameter $\rho > 0$ *such that, for every* $y > 0$,

$$\widehat{H}(y) = \rho \begin{cases} \frac{y-y^\gamma}{1-\gamma} & \text{if } \gamma \neq 1, \\ y \log y & \text{if } \gamma = 1. \end{cases}$$

(iii) If $\gamma \leq 1$ *and* $\eta_* < \infty$, *then there exists a unique solution* $\alpha\colon (0,\infty) \to (0,\infty)$ *to*

$$\frac{\eta\big(t\alpha(t)^{-d}\big)}{t\alpha(t)^{-d}} = \frac{1}{\alpha(t)^2}. \tag{3.42}$$

and it satisfies $\lim_{t\to\infty} t\alpha(t)^{-d} = \infty$. *Moreover,*

(a) If $\gamma = 1$ *and* $0 < \eta_* < \infty$, *then* $\lim_{t\to\infty} \alpha(t) = 1/\sqrt{\eta_*} \in (0,\infty)$.

(b) If $\gamma < 1$ *and* $\eta_* = 0$, *then* $\alpha(t) = t^{\nu+o(1)}$, *where* $\nu = (1-\gamma)/(d+2-d\gamma) \in (0, \frac{1}{d+2}]$.

Certainly, $\alpha(t)$ plays the rôle of the order of the diameter of the intermittent islands for the moments and is identical to the $\alpha(t)$ appearing in Sects. 3.2 and 3.3.

The additional assumption on the convergence of $\eta(t)/t$, which we incorporated in Assumption (H), is mild and is necessary only in the case $\gamma = 1$, which plays a particular rôle.

Now, under Assumption (H), we can formulate a complete distinction into four classes of i.i.d. potentials, in the order from thicker to thinner tails:

(SP) $\eta_* = \infty$ (in particular, $\gamma \geq 1$), the *single-peak case.*
This is the boundary case $\rho = \infty$ of the double-exponential case below. It comprises all heavier-tailed distributions with finite positive exponential moments in the terminology of Example 1.14. We have $\alpha(t) \to 0$ as $t \to \infty$, as is seen from (3.42), i.e., the relevant islands consist of single lattice sites. As we will see in Sect. 6.4.2, this class phenomenologically also contains a number of potentials that have no finite positive exponential moments.

(DE) $\eta_* \in (0,\infty)$ (in particular, $\gamma = 1$), the *double-exponential case.*
This is the case of the double-exponential distribution, see Example 1.12. By rescaling, one can achieve that $\eta_* = 1$. The parameter ρ of Proposition 3.15(ii)(b) is identical to the one in (3.44) below. This case is studied in [GärMol98], [GärHol99], [GärKön00], [GärKönMol00], [GärKönMol07], [BisKön16], [BisKönSan16] and more papers.

(AB) $\eta_* = 0$ and $\gamma = 1$, *the almost bounded case.*

This is the case of islands of slowly growing size, i.e., $\alpha(t) \to \infty$ as $t \to \infty$ slower than any power of t. This case comprises unbounded and bounded from above potentials, see Example 1.13. This class was introduced in [HofKönMör06] and further studied in [GrüKön09]. It lies in the union of the boundary cases $\rho \downarrow 0$ of (DE) and $\gamma \uparrow 1$ of (B).

(B) $\gamma < 1$ (in particular, $\eta_* = 0$), *the bounded case.*
This is the case of islands of rapidly growing size, i.e., $\alpha(t) \to \infty$ as $t \to \infty$ at least as fast as some power of t. Here the potential ξ is necessarily bounded from above. This case was treated for a special subcase of $\gamma = 0$ (Bernoulli traps, see Example 1.10) in [Ant95] and [Ant94], and in generality in [BisKön01] (see Example 1.11).

Let us give some more insight in the four cases. In particular, we give explicit examples for the two characteristic formulas χ of (3.23) and $\chi_\circ$ of (3.32).

Remark 3.16 (The case (SP)) This case is included in [GärMol98] as the upper boundary case $\rho = \infty$ in their notation; it comprises all heavier-tailed potentials with finite positive exponential moments, see Example 1.14, the most explicit examples being the Gaussian distribution and the Weibull distribution $\mathrm{Prob}(\xi(0) > r) = \mathrm{e}^{-Cr^\alpha}$ with $\alpha > 1$. Here $\gamma > 1$, and $\alpha(t) \to 0$, and there is no extinction between the two terms on the left-hand side of (3.26) in Assumption (H). Hence, $H(t)$ and $\eta(t)$ are on the same scale and the subtraction of $yH(t)/\eta(t)$ is trivial. The resulting variational formula is

$$\chi_\circ = \inf_{g \in \ell^2(\mathbb{Z}^d) : \|g\|_2 = 1, g : \mathbb{Z}^d \to \{0,1\}} \|\nabla g\|_2^2 = 2d, \tag{3.43}$$

which is trivially solved exclusively by delta-functions g. This is identical to the value of the right-hand sides of (3.45) and (3.47) for $\rho = \infty$; the corresponding minimisers are $g = \delta_0$ and $\varphi = -\infty \mathbb{1}_{\mathbb{Z}^d \setminus \{0\}}$ (with the understanding that $(-\infty) \cdot 0 = 0$). Actually, [GärMol98] handled this potential class by referring to (3.26) with $\eta(t) = t$ and $\widehat{H}(y) = \infty y \log y$, which also led to $\chi = 2d$, however, with the second term being on the scale t instead of $\eta(t) \approx t^\gamma$. ◇

Remark 3.17 (The case (DE)) The study of this case was initiated in [GärMol98], see also Remark 3.17. The particular interest of this class comes from the fact that the intermittent islands have a discrete and non-trivial structure, since $\alpha(t)$ stays bounded and bounded away from zero and may therefore be put equal to one. The main representative of this class is the *double-exponential distribution* (which is indeed a reflected Gumbel distribution) given by

$$\mathrm{Prob}(\xi(0) > r) = \exp\{-\mathrm{e}^{r/\rho}\}, \qquad r \in \mathbb{R}, \tag{3.44}$$

where $\rho \in (0,\infty)$ is a parameter. Here $H(t) = \rho t \log t + \rho t + o(t)$ as $t \to \infty$. For any representative on the class (DE), $\widehat{H}(y) = \rho y \log y$ and $\eta(t) \sim t$. The characteristic variational problem is given as

$$\chi_\circ = \inf_{g\in\ell^2(\mathbb{Z}^d):\|g\|_2=1} \big[\|\nabla g\|_2^2 + \rho I(g^2)\big], \qquad \text{where } I(g^2) = -\sum_{z\in\mathbb{Z}^d} g^2(z)\log g^2(z). \tag{3.45}$$

Note that one can also write

$$\chi = \inf_{g\in\ell^2(\mathbb{Z}^d):\|g\|_2=1} \Big\langle g^2, \frac{-\Delta^{\rm d} g}{g} - \rho \log g^2\Big\rangle.$$

From here one can easily deduce the Euler-Lagrange equations (i.e., the variational equations that the minimiser solves), which read

$$\frac{-\Delta^{\rm d} g}{g} - \rho \log g^2 = \text{const.} = \chi. \tag{3.46}$$

It is known [GärMol98, GärHol99] that (3.45) possesses minimizers g^2, which are unique (up to spatial shifts) for every sufficiently large ρ (it suffices $\rho > 15.7$). These minimisers are not explicitly known, but they are known to decompose into a d-fold tensor product of minimisers of the formula for $d = 1$, which are positive in $\mathbb{Z}$, have precisely one maximum (assumed to be at zero) and are unimodal (i.e., increasing in $-\mathbb{N}$ and decreasing in $\mathbb{N}$). Furthermore, they approach delta-like functions for $\rho \uparrow \infty$ and Gaussian functions (after rescaling) for $\rho \downarrow 0$. Both asymptotics are consistent with the understanding that (SP) is the boundary case of (DE) for $\rho \uparrow \infty$, and (AB) is the boundary case of (DE) for $\rho \downarrow 0$.

The dual formula for $\chi_\circ$ is

$$\chi = \inf_{\varphi:\mathbb{Z}^d\to\mathbb{R}:\lim_{z\to\infty}\varphi(z)=-\infty} \Big(\frac{\rho}{\rm e}\sum_{z\in\mathbb{Z}^d} {\rm e}^{\varphi(z)/\rho} - \lambda(\varphi)\Big), \tag{3.47}$$

where $\lambda(\varphi) = \sup_{g\in\ell^2(\mathbb{Z}^d):\|g\|_2=1}\langle g, (\Delta^{\rm d} + \varphi)g\rangle$ denotes the top of the spectrum of $\Delta^{\rm d} + \varphi$ in $\mathbb{Z}^d$; note that, due to the condition $\lim_{z\to\infty}\varphi(z) = -\infty$, it is also its principal eigenvalue with (up to shift and normalisation) precisely one eigenfunction. The first term in (3.47) is easily seen from (3.44) to be the infinite-space version of the large-deviation rate function of the potential; obviously the condition $\lim_{z\to\infty}\varphi(z) = -\infty$ is necessary for it to be finite.

Both formulas, $\chi_\circ$ and χ have been proved to be stable in the sense that every sequence of approximate minimisers has a subsequence that converges pointwise to some shift of the minimiser, see [GärKönMol07]. ◇

Remark 3.18 (The case (AB)) This class was brought to the surface in [HofKönMör06] and was further studied in [GrüKön09]; it is a kind of interpolation between the cases (DE) for $\rho \approx 0$ and (B) for $\gamma \approx 1$, see also Remark 3.18. One obtains examples of potentials (unbounded from above) by replacing ρ in (3.44) by a sufficiently regular function $\rho(r)$ that tends to 0 as $r \to \infty$, and other examples (bounded from above) by replacing γ in (3.51) by a sufficiently regular function $\gamma(x)$ tending to 1 as $x \downarrow 0$. We find that $\widehat{H}(y) = \text{const}\, y \log y$, and the infinite-space version of the rate function in (3.14) is

$$I(\varphi) = \frac{\rho}{\mathrm{e}} \int_{\mathbb{R}^d} \mathrm{e}^{\varphi(x)/\rho}\, \mathrm{d}x. \tag{3.48}$$

The characteristic variational problem is given by

$$\chi_\circ = \inf_{g \in H^1(\mathbb{R}^d): \|g\|_2 = 1} \Big[\|\nabla g\|_2^2 + \rho \int_{\mathbb{R}^d} g^2 \log g^2 \Big]. \tag{3.49}$$

This is the spatially continuous version of (3.45), but in contrast it does admit explicit minimisers. They are easily seen to be (up to spatial shifts, uniquely) equal to the Gaussian density $g^2(x) = \text{const}\, \mathrm{e}^{-\rho\|x\|_2^2}$, which is the principal eigenfunction of $\Delta + \varphi$ for the parabolic function $\varphi(x) = \text{const} - \rho^2 \|x\|_2^2$. The parabola in turn is the (up to spatial shifts, unique) minimiser of the alternate representation of χ:

$$\chi = \inf_{\varphi \in \mathcal{C}(\mathbb{R}^d): \lim_{x\to\infty} \varphi(x) = -\infty} \Big(\frac{\rho}{\mathrm{e}} \int_{\mathbb{R}^d} \mathrm{e}^{\varphi(x)/\rho}\, \mathrm{d}x - \lambda(\varphi) \Big), \tag{3.50}$$

where $\lambda(\varphi) = \sup_{g \in H^1(\mathbb{R}^d): \|g\|_2 = 1} \langle g, (\Delta + \varphi) g \rangle$ is the principal eigenvalue of the operator $\Delta + \varphi$ in $L^2(\mathbb{R}^d)$. Hence, in spite of a relatively odd definition of the potential distribution, the appropriately rescaled and shifted shape of the local times and of the potential that give the main contribution to the moments of the total mass are unique, explicit and elementary functions. Using a standard logarithmic Sobolev inequality, some stability properties of both formulas, $\chi_\circ$ and χ can be derived, see [GrüKön09]. ◇

Remark 3.19 (The case (B)) This class contains only distributions that are bounded from above, so without loss of generality we assume that their essential supremum is equal to zero. The upper tails at zero of the main representatives are given by

$$\log \text{Prob}(\xi(0) > -x) \sim -D x^{-\frac{\gamma}{1-\gamma}}, \qquad x \downarrow 0, \tag{3.51}$$

where $D \in (0, \infty)$ and $\gamma \in [0, 1)$ are parameters (certainly, γ is identical with the γ of Proposition 3.15). Let us first turn to the case $\gamma \in (0, 1)$. We find that $H(t) = -t^{\gamma + o(1)}$ for $t \to \infty$ and therefore $\widehat{H}(y) = \frac{\rho}{\gamma - 1}(y^\gamma - y)$ for some $\rho \in (0, \infty)$ that depends on D. Hence,

$$\chi_\circ = \inf_{g \in H^1(\mathbb{R}^d): \|g\|_2 = 1} \Big(\|\nabla g\|_2^2 - \rho \int_{\mathbb{R}^d} \frac{g^{2\gamma} - g^2}{\gamma - 1} \Big). \tag{3.52}$$

This formula was analysed in [Sch11]. It is in particular shown that a minimiser exists, is unique up to spatial shifts, and has compact support, which is actually a ball. Certainly, the last term $\rho g^2/(\gamma-1)$ can be easily extracted from the integral, as it gives just the constant $\rho/(\gamma-1)$. Actually in [BisKön01], the second term on the left-hand side of (3.26) in Assumption (H) (i.e., the term $yH(t)/\eta(t)$) was dropped, as the first one alone behaves like a constant times t^γ, and hence the last term in (3.52) did not appear. However, the formula in (3.52) has the advantage that one can easily guess the limit as $\gamma \uparrow 1$. Indeed it is not too difficult to identify this limit as (3.49), as, obviously, the term $\frac{g^{2\gamma}-g^2}{\gamma-1}$ converges toward the derivative of $\gamma \mapsto (g^2)^\gamma$ at $\gamma = 1$, which is equal to $g^2 \log g^2$.

The case $\gamma = 0$ contains the case of Bernoulli traps, where only the values 0 and $-\infty$ are attained, the former with probability e^{-D}. For all distributions in the case $\gamma = 0$, we also have $H(t) = t^{o(1)}$ and therefore $\widehat{H}(y) = D\mathbb{1}_{(0,\infty)}(y)$. Hence,

$$\chi_\circ = \inf_{g \in H^1(\mathbb{R}^d) : \|g\|_2 = 1} \Big(\|\nabla g\|_2^2 + D|\mathrm{supp}\,(g)| \Big). \tag{3.53}$$

This is a classic variational formula that is well understood for a long time. Again, a minimiser exists, is unique up to spatial shifts, and has compact support, which is actually a ball. The ball-shape of the support follows from a classic isoperimetric inequality called the *Faber-Krahn inequality* (see [Ban80], e.g.), which says that, among all regular domains with a given finite volume, the principal Dirichlet eigenvalue of $-\Delta$ in that domain is minimal precisely for a ball. Nevertheless, this does not imply stability of the formula, even though it holds; see Sect. 7.1.

The minimiser can be explicitly written in terms of the principal eigenfunction of the Laplace operator in a ball, and the radius of that ball can easily be found using elementary analysis. Indeed, use that $\lambda(rQ) = r^{-2}\lambda(Q)$ for any $r > 0$ and any compact set $Q \subset \mathbb{R}^d$, where we wrote now $\lambda(A)$ for the principal Dirichlet eigenvalue of the continuous Laplace operator Δ in the set A. Then we see easily that, writing B_r for the centred ball with radius r and ω_d for the volume of B_1,

$$\begin{aligned}\chi_\circ &= \min_{A \subset \mathbb{R}^d} (\lambda(A) + D|A|) = \min_{r \in (0,\infty)} \big(\lambda(B_r) + D|B_r|\big) = \min_{r \in (0,\infty)} \big(r^{-2}\lambda(B_1) + Dr^d\omega_d\big) \\ &= \Big(D\omega_d\big(\tfrac{2\lambda(B_1)}{d}\big)^d\Big)^{\frac{2}{d+2}} \big(1 + \tfrac{d}{2\lambda(B_1)}\big),\end{aligned} \tag{3.54}$$

and the optimal r is $r^* = (\frac{2}{\lambda(B_1)} D\omega_d d)^{1/(d+2)}$. Let us also remark that the value of $\lambda(B_1)$ can be expressed in terms of the smallest zero of a Bessel function, see [Szn98], e.g. ◇

Remark 3.20 (Moment intermittency) Let us remark that one can easily verify on a case-by-case basis that in all the preceding cases moment intermittency holds in the sense that (1.10) holds. Also part of the geometric interpretation of intermittency was already clear from the heuristics of the moment asymptotics (believing that a corresponding upper bound holds), as the total mass was well approximated by the sub-sum over the ball with radius $R\alpha(t)$ only. This shows that this ball is the (unique) intermittent island in the annealed setting. However, it was not even attempted to show that the sum from outside this ball gives a negligible contribution. This is indeed a rather difficult task to do, which has been done only in the case (DE), which we report on in Sect. 7.1. Such an assertion cannot be proved by looking at the leading two terms only, since a concentration of the mass in a shift of the island even by some quite large amount contributes the same leading two terms. ◇

Remark 3.21 (Lifshitz tails for bounded potentials) As announced in Sect. 2.2.6, we give an example of an assertion about the Lifshitz tails for the random Schrödinger operator $\Delta^{\mathrm{d}} + \xi$ for an i.i.d. potential ξ with single-site distribution given in (3.51). From (3.38) and Remark 3.19, we have, for any $p \in (0, \infty)$,

$$\log\langle U(t)^p\rangle \sim \frac{pt}{\alpha(pt)^2}\widehat{\chi}_\circ, \qquad t \to \infty,$$

where $\alpha(t) = t^{\nu+o(1)}$ with $\nu = \frac{1-\gamma}{d+2-\gamma} \in (0, \frac{1}{d+2}]$, and $\widehat{\chi}_\circ = \chi_\circ + \rho/(\gamma - 1)$ with $\chi_\circ$ as in (3.52). According to (2.29), we have the same asymptotics for the Laplace transform $\mathcal{L}(N, t)$ of the IDS N with parameter t. Inverting this transform, we obtain [BisKön01, Theorem 1.3]

$$\lim_{E\downarrow 0} \frac{\log N(E)}{E\alpha^{-1}(E^{-1/2})} = -\frac{2\nu}{1-2\nu}\big[(1-2\nu)\widehat{\chi}_\circ\big]^{1/2\nu},$$

where α^{-1} is the inverse function of $t \mapsto \alpha(t)$. As a consequence, we have $\log N(E) \asymp E^{1/\beta+o(1)}$ for $E \downarrow 0$ with Lifshitz exponent $\beta = \frac{2\nu}{1-2\nu} \in (0, \frac{2}{d}]$. This means that one can obtain any Lifshitz exponent in $(0, \frac{2}{d}]$ by making the tails of the single-site potential sufficiently thin, the boundary case of hard obstacles (i.e., $\gamma = 0$) being the boundary case with $\beta = \frac{2}{d}$. ◇

3.5 The Spatially Continuous Case

In the spatially discrete case, we formulate the main assumption on the distribution of the random potential in terms of just one single random variable $\xi(0)$, since we rely on the assumption that the potential is i.i.d. In this way, one naturally covers all i.i.d. potentials whose single-site distribution is regular in the sense that Assumption (H) holds.

However, in the continuous case, one cannot do this so easily without determining the spatial correlations (if we do not want to mimic the i.i.d. case by putting the potential constant in the unit boxes $z+(-\frac{1}{2}, \frac{1}{2}]^d$ and i.i.d. over $z \in \mathbb{Z}^d$). Nevertheless, a number of potentials studied in the literature obviously belong to one of the above classes in a phenomenological sense. E.g., the case of a Poisson field of obstacles and many variants belong to the case (B), see Sect. 3.5.1.

However, new variational formulas arise in the cases of regular Gaussian fields and Poisson shot-noise fields with positive cloud, see Sects. 3.5.2 and 3.5.3. Here the radius of the intermittent islands does asymptotically shrink to zero, but after rescaling an interesting, smooth shape is developed in both the random potential and in the solution; a phenomenon that is not possible in discrete space. Hence, these two important examples do not belong to the class (SP). In both cases, the first-order term of the moment asymptotics turns out to be given by $\mathrm{e}^{H(t)} = \langle \mathrm{e}^{tV(0)} \rangle$ (where we recall that we write V for the random potential), like in the spatially discrete case.

3.5.1 *Brownian Motion Among Poisson Obstacles*

Let us here discuss the moment asymptotics in the important case of a Brownian motion among Poisson obstacles, see Example 1.15. This was studied in [DonVar75], in many papers by Sznitman, resulting in the monograph [Szn98], and in more papers. We assume for simplicity that the potential is given as the hard-trap potential, i.e., $V(x) = -\sum_{i\in\mathbb{N}} W(x - x_i)$ with $(x_i)_{i\in\mathbb{N}}$ a standard Poisson point process on $\mathbb{R}^d$ with intensity ν and $W = \infty \times \mathbb{1}_{K_a}$, where K_a is the centred ball with radius $a > 0$. In general, the cloud $W: \mathbb{R}^d \to [0, \infty)$ is assumed bounded, measurable and compactly supported (e.g., an indicator function on a compact polar set), but the following asymptotics are shown in [DonVar75] to hold literally true also for clouds like $W(x) = C|x|^{-q}$ with $q \in (d + 2, \infty)$. The interesting case $q \in (d, d+2)$ was shown in [Fuk11] to show a different behaviour, see Example 7.6 below. The case $q \in (d/2, d)$ requires a normalisation and exhibits further new phenomena [CheKul11, Che12, CheKul12], see Sect. 7.3.4.

We want to understand the large-t asymptotics of the moments of

$$U(t) = \mathbb{E}_0\Big[\exp\Big\{-\int_0^t \sum_{i\in\mathbb{N}} W(Z_s - x_i)\,\mathrm{d}s\Big\}\Big], \tag{3.55}$$

where $Z = (Z_s)_{s\in[0,\infty)}$ is the Brownian motion with generator Δ (i.e., the time-change of the standard Brownian motion with a factor of 2). Let us first take the expectation with respect to the Poisson process. We recall from Example 1.15 that, in the special case $W = \infty \mathbb{1}_{K_a}$,

$$\langle U(t) \rangle = \mathbb{E}_0[\mathrm{e}^{-\nu|S_a(t)|}],$$

where $S_a(t) = \bigcup_{s\in[0,t]} K_a(Z_s)$ is the Wiener sausage up to time t with radius a. For finding the large-t asymptotics of this (sometimes called the *Wiener sausage problem*), applying a large-deviation principle (LDP) for the normalised occupation times of the motion, $\mu_t = \frac{1}{t}\int_0^t \delta_{Z_s}\,\mathrm{d}s$ is the most natural method. This LDP roughly says that

$$\mathbb{P}_0\Big(\mu_t(\mathrm{d}x) \approx \phi^2(x)\,\mathrm{d}x\Big) \approx \exp\Big\{-t\|\nabla\phi\|_2^2)\Big\}, \qquad t \to \infty, \tag{3.56}$$

i.e., the probability that μ_t resembles the probability measure with density ϕ^2 decays exponentially fast with rate given by the energy of ϕ. For a precise statement, one has to restrict to finite boxes and has to consider topologies. See Sect. 4.2 below for precise statements and [DemZei98] for more about the theory.

The LDP in (3.56) describes the behaviour of the motion when staying by time t in a compact part of $\mathbb{R}^d$, which does not depend on t. However, the typical spatial scale of the motion in the expectation of $\mathrm{e}^{-\nu|S_a(t)|}$ is not the finite one. Rather, we seek for some scale α_t such that a compromise between the probability to stay in a box of radius α_t and the term $\mathrm{e}^{-\nu|S_a(t)|}$ is realised. Looking only at the exponential rates, the rate of the former is $t\alpha_t^{-2}$, which can easily be seen from the argument that leads to (2.30), and the scale of the latter is α_t^d, both with the negative sign. Minimising their sum shows that the optimal scale is $\alpha_t = t^{1/(d+2)}$. Due to the Brownian scaling property for this spatial scale, we immediately obtain from (3.56) an LDP for the normalised occupation measure $\mu_t^{(t)}$ of the rescaling $Z_s^{(t)} = \alpha_t^{-1}Z_{s\alpha_t^2}$; indeed, it is the same with scale $t\alpha_t^{-2} = t^{d/(d+2)}$ instead of t. For large t, we may neglect the radius a of the Wiener sausage and can approximate

$$|S_a(t)| \approx \alpha_t^d|\mathrm{supp}\,(\mu_t^{(t)})| = t^{d/(d+2)}|\mathrm{supp}\,(\mu_t^{(t)})|. \tag{3.57}$$

Hence, using Varadhan's lemma (see Sect. 4.2 for a precise formulation), we now understand that

$$\langle U(t)\rangle = \mathbb{E}_0[\mathrm{e}^{-\nu|S_a(t)|}] \approx \mathbb{E}_0\big[\mathrm{e}^{-\nu t^{d/(d+2)}|\mathrm{supp}\,(\mu_t^{(t)})|}\big] \approx \mathrm{e}^{-t^{d/(d+2)}\chi_\circ}, \tag{3.58}$$

where

$$\chi_\circ = \inf_{g\in H^1(\mathbb{R}^d):\|g\|_2=1}\big(\|\nabla g\|^2 + \nu|\mathrm{supp}\,(g^2)|\big). \tag{3.59}$$

(The '$\approx$' in (3.58) means that the error is $\mathrm{e}^{o(t^{d/(d+2)})}$.) Note that $\chi_\circ$ is identical with the $\chi_\circ$ from (3.53) with $D = \nu$. This ends our heuristic explanation of the asymptotics of the moments in the case of Brownian motion among Poisson obstacles, using large deviations for the motion, analogously to Sect. 3.3. From the point on that we approximated with the Brownian scaling $(\alpha_t^{-1}Z_{s\alpha_t^2})_{s\in[0,\infty)}$, the range problem for Brownian motion is essentially analogous to the spatially discrete case, i.e., to the case of Bernoulli traps in case (B).

Another way to understand the asymptotics can be heuristically described using a joint strategy of the Poisson process and the motion as follows (this is analogous to Sect. 3.2). We start from (3.55) and observe that the large-t asymptotics of $\langle U(t)\rangle$

are mainly concentrated on maximisation of the term in the exponent, i.e., on minimisation of $\int_0^t \sum_{i\in\mathbb{N}} W(Z_s - x_i)$. Let us assume that W is a bounded function with compact support. One good joint strategy of the motion Z and the point process $\omega = (x_i)_i$ is that the latter leaves a certain bounded area $A \subset \mathbb{R}^d$ empty of sites x_i (even with a certain distance, such that A does not intersect any support of the functions $W(\cdot - x_i)$, but this extra amount asymptotically vanishes), and the path $Z_{[0,t]}$ does not leave A. In this case, the exponent is even equal to zero, which is certainly optimal. The probability cost for ω of following this strategy is $\mathrm{e}^{-\nu|A|}$, the Poisson probability to have no particle in a set with Lebesgue measure $|A|$. The cost for the motion can be found from the LDP of (3.56) as follows:

$$\mathbb{P}_0(Z_{[0,t]} \subset A) = \mathbb{P}_0(\mathrm{supp}\,(\mu_t) \subset A) \approx \exp\Big\{ -t \inf_{g\in H_0^1(A):\|g\|_2=1} \|\nabla g\|_2^2 \Big\}.$$

Note that the Rayleigh-Ritz principle (a continuous version of (2.20)) says that the right-hand side is nothing but $\mathrm{e}^{-t\lambda(A)}$, where $\lambda(A)$ is the principal eigenvalue of Δ in A with zero boundary condition. Hence, the joint strategy has the probabilistic cost $\mathrm{e}^{-(\nu|A|+t\lambda(A))}$, and this is roughly equal to the expectation of $U(t)$. Now we have to identify the optimal set A, i.e., to identify the minimiser A in the formula

$$\chi_t = \inf_{A\subset\mathbb{R}^d} (\nu|A| + t\lambda(A)).$$

Using elementary scaling arguments, it is easily seen that the diameter of the optimal A is of order $t^{1/(d+2)}$ and that χ_t is of order $t^{d/(d+2)}$ and that $\chi_t t^{-d/(d+2)}$ converges towards $\chi_\circ$ in the form of (3.54) with $D = \nu$. This ends the second derivation of the moment asymptotics.

These asymptotics do not depend on the shape of the cloud function W, at least as long as it has a compact support and does not depend on t. Actually, to a certain degree, they are stable with respect to multiplying W with a t-dependent factor; see Sect. 7.3.3 for the boundaries of this.

Remark 3.22 (Gibbsian point fields) In [Szn93], the potential $V(x) = -\sum_{i\in\mathbb{N}} W(x - x_i)$ is considered with $(x_i)_i$ being a Gibbsian point field (see Example 1.17), i.e., an ergodic point process with correlation between the particles induced by a symmetric pair potential, which is assumed to be non-negative, bounded from below, compactly supported and superstable. The cloud function W is taken as the hard-core function $W = \infty \mathbb{1}_{K_a}$ there, where K_a is the centred ball with radius a. Under these assumptions, the results for the moments of the total mass (and also for its almost sure behaviour; see Remark 5.12) are literally identical with the above result; however the proofs are more involved. ◇

3.5.2 Gaussian Potentials

Also the case of a Gaussian potential is interesting. Let $V = (V(x))_{x\in\mathbb{R}^d}$ be a Hölder continuous stationary centred Gaussian field with covariance function $B(x) = \langle V(0)V(x)\rangle$. We assume that B is twice continuously differentiable in a neighbourhood of zero with $B(0) = \sigma^2 \in (0,\infty)$ and such that $-B''(0) = \Sigma^2$ is a positive definite matrix, i.e., the maximum of B at zero is strict and B parabola-shaped. Then $H(t) = \log\langle e^{tV(0)}\rangle = \frac{1}{2}t^2\sigma^2$, and with $\alpha(t) = t^{-1/4}$, it is proved in [GärKön00] that

$$\langle U(t)\rangle = e^{\frac{1}{2}t^2\sigma^2}\exp\Big\{-\frac{t}{\alpha(t)^2}(2^{-1/2}\mathrm{tr}(\Sigma)+o(1))\Big\}, \qquad t\to\infty, \tag{3.60}$$

where tr denotes the trace. The enormously explicit term in the second-order term in fact comes from a variational formula that is in the spirit of the characteristic formula (3.32); this is explained in Example 7.4. Like for the formula (3.49) in the class (AB), the minimisers are perfect Gaussian functions (describing the rescaled local times), and the corresponding potential shapes are perfect parabolas, like the second-order approximation of covariance function B close to zero. The scale $\alpha(t)$ has the same interpretation as in the above heuristics as the order of the radius of the relevant islands. Interestingly, this is an example for $\alpha(t)\to 0$, i.e., the Gaussian field attains the relevant maxima on very small islands. Such an interesting peak behaviour on vanishing islands can be observed only in the spatially continuous case. The first term in the above asymptotics was already derived in [CarMol95].

Gaussian potentials with much less regularity and the singularity $B(0)=\infty$ and $B(x)\sim |x|^{-\gamma}$ as $x\to 0$ for some $\gamma\in(0,2)$ are considered in [Che14], see Sect. 5.13. The case of Gaussian white noise is widely open, see Example 1.21.

3.5.3 Poisson Shot-Noise Potential with High Peaks

As in Sect. 3.5.1 and Example 1.16, let $(x_i)_{i\in\mathbb{N}}$ be a standard Poisson point process on $\mathbb{R}^d$ with intensity $\nu\in(0,\infty)$ and $\varphi\colon\mathbb{R}^d\to[0,\infty)$ be a non-negative, compactly supported cloud, and we consider the potential $V(x)=\sum_{i\in\mathbb{N}}\varphi(x-x_i)$. Like for the covariance function in Sect. 3.5.2, we assume that φ is strictly maximal in 0 with a strictly positive definite Hessian matrix $\Sigma^2 = -\varphi''(0)$. Clearly, the logarithmic moment generating function $H(t)=\log\langle e^{tV(0)}\rangle$ is given by $H(t)=\nu\int(e^{t\varphi(x)}-1)\,dx$. Then in [GärKön00] it turns out that the order of the diameter of the relevant islands is given as $\alpha(t)=t^{d/8}e^{-t\varphi(0)/4}$, and that

$$\langle U(t)\rangle = e^{H(t)}\exp\Big\{-\frac{t}{\alpha(t)^2}\Big(\Big(\nu\frac{(2\pi)^{d/2}}{2\det(\Sigma)}\Big)^{1/2}\mathrm{tr}(\Sigma)+o(1)\Big)\Big\}, \qquad t\to\infty. \tag{3.61}$$

Note the extremely strong decay of $\alpha(t)$ and the extremely fast asymptotics of the moments of the total mass. The first term, $H(t)$, seems to depend on all the values of the cloud φ in a neighbourhood of zero, but an application of the Laplace method shows that its main asymptotics depend only on $\varphi(0)$ and the Hessian matrix of φ at zero. However, the second term depends only on the Hessian matrix at zero.

Interestingly, both (3.60) and (3.61) may be summarized as

$$\frac{1}{t}\log\langle U(t)\rangle = \frac{H(t)}{t} - (\chi + o(1))\sqrt{H'(t)}, \qquad t \to \infty, \tag{3.62}$$

with $\chi = (2\sigma^2)^{-1/2}\mathrm{tr}(\Sigma)$, where $\sigma^2 = \varphi(0)$ in the Poisson case. These two interesting cases were examined as special cases of the PAM with a correlated random potential that admits a natural large-deviations approach, which we explain in Example 7.4.

Chapter 4
Some Proof Techniques

In Sect. 3.2, we described the heuristics for the large-t exponential rate of the moments of the total mass $U(t)$, neglecting all technical issues. More precisely, we only gave the heuristics for a lower bound and assured that this matches with the upper bound, at least on the level of precision that we discussed, i.e., logarithmic asymptotics on the scale $t/\alpha(t)^2$.

However, in the proofs of the upper bound there are a number of problems to be solved, which require quite some efforts and developments. These problems may make a technical impression, as large-deviation theory already quite clearly suggests the validity of the asymptotics (as soon as one is willing to believe that regularity and compactness issues are only technical). However, the mathematical difficulties in deriving proper proofs are not to be underestimated. Indeed, a number of papers have been written essentially with the main purpose to overcome them by introducing new methods. These problems served as motivations for finding them, as one felt that these new methods might be applicable to other interesting situations.

In this chapter, we present some of the proof techniques that have been successfully used in this respect. In Sects. 4.2, 4.3, 4.4, 4.5, 4.6, 4.7, 4.8, 4.9, 4.10, and 4.11 we explain the methods in their simplest form, their nature, benefits and drawbacks. Let us first describe in Sect. 4.1 the type of problems that we want to solve.

4.1 The Problems

One type of crucial assertions are precise logarithmic large-t upper bounds for expressions of the form

$$\mathbb{E}_0\Big[\exp\Big\{\gamma_t \int_{\mathbb{R}^d} \widehat{H}(L_t(x))\,\mathrm{d}x\Big\}\Big], \tag{4.1}$$

W. König, *The Parabolic Anderson Model*, Pathways in Mathematics,
DOI 10.1007/978-3-319-33596-4_4

where $\gamma_t = t\alpha(t)^{-2} \to \infty$ is a scale function, $\widehat{H}$ is the function introduced in Assumption (H), see (3.26), and L_t is the rescaled and normalized local times introduced in (3.28). We stick here to the case $\alpha(t) \to \infty$, i.e., the case where a spatial scaling is necessary, but some of the following applies also to $\alpha(t) \equiv 1$ with the integral on $\mathbb{R}^d$ replaced by a sum on $\mathbb{Z}^d$ and L_t replaced by $\frac{1}{t}\ell_t$ and is technically even easier. Recall from Sect. 3.3 that (4.1) comes from the expectation of $U(t)$, after having taken the expectation with respect to the random potential and inserting some rescaling into the local times of the random walk. The main goal is to prove that the negative exponential rate on the scale γ_t is equal to $\chi_\circ$ defined in (3.32), i.e., to the infimum of the LDP rate function for the local times minus the functional $g^2 \mapsto \int \widehat{H} \circ g^2$, taken over all L^2-normalized functions $g \in H^1(\mathbb{R}^d)$. Another fundamental and closely related task is to find a tight logarithmic upper bound for the expression

$$\left\langle \mathrm{e}^{t\lambda^{\mathrm{d}}(B,\xi)} \right\rangle, \tag{4.2}$$

where $\lambda^{\mathrm{d}}(B,\xi)$ is the principal Dirichlet eigenvalue of the operator $\Delta^{\mathrm{d}} + \xi$ in the box $B \subset \mathbb{Z}^d$, whose radius may depend on t and may be rather large. Here we did not yet insert the $\alpha(t)$-depending rescaling of the eigenvalue that proved necessary in our heuristic derivation in Sect. 3.2.

As we have seen in the preceding sections, in particular see (2.21) and (2.22), both tasks are strongly related, and there are techniques to estimate the two expressions in (4.1) and (4.2) in terms of each other.

However, a number of technical obstacles arise in deriving a tight logarithmic upper bound for (4.1) and (4.2), respectively:

(1) restriction of the integral $\int_{\mathbb{R}^d}$ respectively of the box B to a box of appropriate (much smaller) size,
(2) overcoming the lack of boundedness of $\widehat{H}$, respectively of the map $\xi \mapsto \lambda^{\mathrm{d}}(B,\xi)$ and
(3) overcoming the lack of continuity of $\widehat{H}$, respectively of the map $\xi \mapsto \lambda^{\mathrm{d}}(B,\xi)$.

These are problems that are similar also to those that arise in the analysis of other exponential functionals of the local times (for example self-intersection local times, where $\widehat{H}(l) = l^p$ for some $p > 1$, see Example 7.9, Sect. 7.3.2, the monograph [Che10] and the short survey [Kön10]).

In Sect. 4.2 we first explain why the desired result should be true at all, and why items (1)–(3) are indeed problems.

4.2 Large Deviations

One of the cornerstones of the mathematical analysis of the expectation of exponential functionals with a large prefactor is the theory of large deviations, see [DemZei98] for a comprehensive treatment. A family $(Y_t)_{t\in(0,\infty)}$ of random variables with values in some topological space $\mathcal{X}$ is said to *satisfy a large-deviation principle (LDP)* with *speed* γ_t and *rate function* $I: \mathcal{X} \to [0, \infty]$ if the level sets $\{x \in \mathcal{X}: I(x) \le c\}$ are compact for any $c \in \mathbb{R}$ and if the set functions $\frac{1}{\gamma_t} \log \mathbb{P}(Y_t \in \cdot)$ converge weakly towards the set function $A \mapsto \inf_A I = \inf_{x \in A} I(x)$ in the sense that

$$\limsup_{t\to\infty} \frac{1}{\gamma_t} \log \mathbb{P}(Y_t \in \mathcal{C}) \le -\inf_{\mathcal{C}} I \qquad \text{and} \qquad \liminf_{t\to\infty} \frac{1}{\gamma_t} \log \mathbb{P}(Y_t \in \mathcal{O}) \ge -\inf_{\mathcal{O}} I$$

for any closed set $\mathcal{C}$ and any open set $\mathcal{O}$ in $\mathcal{X}$, respectively. This – somewhat technical – definition may be symbolically summarized by saying that $\mathbb{P}(Y_t = x) \approx \mathrm{e}^{-\gamma_t I(x)}$.

One of the most important ideas is the following strong extension of the well-known Laplace method, which is called *Varadhan's lemma*: If $(Y_t)_{t \in (0,\infty)}$ satisfies the above LDP and $F: \mathcal{X} \to \mathbb{R}$ is a continuous and bounded function, then

$$\lim_{t\to\infty} \frac{1}{\gamma_t} \log \mathbb{E}\Big[\mathrm{e}^{\gamma_t F(Y_t)}\Big] = \sup_{\mathcal{X}}(F - I).$$

4.2.1 LDPs for the Occupation Measures of Random Motions

One could already guess from the representation of the moments of $U(t)$ in (2.14), and it has been used in the heuristics in Sect. 3.3, that the analysis of the moments of $U(t)$ may be very well attacked with the help of an LDP for the normalised local times $\frac{1}{t}\ell_t$ or spatially rescaled versions of them, see (3.30), of the underlying simple random walk $(X_s)_{s \in [0,\infty)}$. Let us cite the relevant LDPs from [Gär77] and [DonVar75-83]. By $\mathcal{M}_1(B)$ we denote the set of probability measures on $\mathbb{Z}^d$ with support in $B \subset \mathbb{Z}^d$.

Lemma 4.1 (LDP for the local times of the random walk) *For any finite box $B \subset \mathbb{Z}^d$, the normalised local times $\frac{1}{t}\ell_t = \frac{1}{t}\int_0^t \delta_{X_s}\,\mathrm{d}s$ satisfy an LDP with speed t both under the distribution of the random walk conditioned on not exiting B by time t and under the distribution of the periodised random walk. The rate function of the former is the quadratic form*

$$\mathcal{M}_1(B) \ni \mu \mapsto \big\langle \sqrt{\mu}, -\Delta^{\mathrm{d}} \sqrt{\mu} \big\rangle - C_B = \sum_{x,y \in \mathbb{Z}^d : x \sim y} \Big(\sqrt{\mu_x} - \sqrt{\mu_y}\Big)^2 - C_B,$$

where $C_B = \inf_{\mu \in \mathcal{M}_1(\mathbb{Z}^d): \mu(B)=1} \langle \sqrt{\mu}, -\Delta^{\mathrm{d}} \sqrt{\mu} \rangle > 0$. The rate function of the latter is the analogous quadratic form with Δ^{d} replaced by the discrete Laplace operator on B with periodised boundary condition in B and C_B replaced by 0.

The zero boundary case is true for any finite set B, not necessarily a box.

Remark 4.2 **(C_B and eigenvalues)** The normalisation constant C_B comes from the exponential rate of the probability for the conditioning:

$$
\begin{aligned}
C_B &= -\lim_{t\to\infty}\frac{1}{t}\log\mathbb{P}_0(X_s\in B \text{ for every } s\in[0,t]) \\
&= -\lim_{t\to\infty}\frac{1}{t}\log\mathbb{P}_0\big(\operatorname{supp}(\tfrac{1}{t}\ell_t)\subset B\big).
\end{aligned}
$$

On the other hand, one sees from comparing to the Rayleigh-Ritz formula in (2.20) that C_B is equal to the principal eigenvalue of Δ^{d} with zero boundary condition in B. When considering zero boundary condition, it is positive, and when considering periodised boundary condition it is equal to zero, as the corresponding eigenfunction is constant. ◇

The two LDPs of Lemma 4.1 are important tools for the case of the double-exponential distribution (i.e., the cases(DE) and (SP) in [GärMol98], where $\alpha(t)=1$ respectively $\alpha(t)\to 0$, i.e, in the absence of spatial rescaling. Note that, for time-discrete random walk, there is also a LDP like the one in Lemma 4.1, but the rate function is different, as this LDP is based on an LDP for the empirical *pair* measures via the contraction principle.

However, in the cases (AB) and (B), we need to consider the spatially rescaled version L_t of ℓ_t introduced in (3.28). A proper formulation of (3.30) is as follows; see [GanKönShi07, Lemma 3.1] for the discrete-time case, but an extension to the continuous-time case is simple, see [HofKönMör06, Prop. 3.4].

Lemma 4.3 (LDP for the rescaled local times of the random walk) *Assume that $\alpha(t)\to\infty$ such that $\alpha(t)\ll\sqrt{t}$ in $d=1$, $\alpha(t)^d\ll t/\log t$ in $d=2$ and $\alpha(t)^d\ll t$ in $d\geq 3$. Then, for any centred cube $Q\subset\mathbb{R}^d$, the rescaled local times $(L_t)_{t\in(0,\infty)}$, both under the distribution of the random walk conditioned on not exiting $(\alpha(t)Q)\cap\mathbb{Z}^d$ by time t and under the distribution of the periodised random walk in that box, satisfies an LDP on the set of probability densities on Q with speed $t\alpha(t)^{-2}$ and rate function*

$$
g^2\mapsto\begin{cases}\|\nabla g\|_2^2-\lambda(Q) & \text{if } g\in H_0^1(Q),\\ +\infty & \text{otherwise,}\end{cases} \tag{4.3}
$$

in the case of zero boundary condition, and $g^2\mapsto\|\nabla g\|_2^2$ in the case of periodic boundary condition. The topology is the one that is induced by test integrals with respect to continuous functions $Q\to\mathbb{R}$.

Here $H_0^1(Q)$ is the usual Sobolev space of L^2-functions $g\colon\mathbb{R}^d\to\mathbb{R}$ that possess a gradient in the weak sense with zero boundary condition in Q; it is usually defined as the completion of the set of all infinitely smooth functions $g\colon\mathbb{R}^d\to\mathbb{R}$ with support in Q_R with respect to the L^2-norm of g plus the one of ∇g. The normalisation $\lambda(Q)$ is equal to the principal eigenvalue of Δ in Q with zero respectively periodic boundary condition.

In (3.56), we used a similar LDP with speed equal to t for the normalised occupation times measures of the Brownian motion with generator Δ; this LDP holds for the two cases (1) under conditioning the motion not to leave the set Q by time t and (2) for periodised Brownian motion in $[-R, R]^d$. Both LDPs also follow from [Gär77] and [DonVar75-83], and the rate function is again equal to the function in (4.3) with zero respectively with periodic boundary condition. We call this LDP sometimes the *Donsker-Varadhan-Gärtner LDP*.

The fundamental papers [DonVar75] and [DonVar79] by Donsker and Varadhan on the Wiener sausage contain apparently the first substantial annealed results on the asymptotics for the PAM, based on the large-deviation theory that they developed in [DonVar75-83] and periodisation (see Sect. 4.3).

4.2.2 LDPs for the Random Potential

In Sect. 3.2 we used an LDP for the shifted and rescaled potential $\overline{\xi}_t$ in (3.19), based on the Assumption (J) for the upper tails of a single-site potential variable, see (3.20). Deriving precise versions of such an LDP in some topology presents no deep problems, in particular in the simpler case $\alpha(t) = 1$, where no spatial scaling is involved. See [BenKipLan95] for such assertions and proofs. However, it is notoriously difficult to complete the main step in the proof, the argument for the application of Varadhan's lemma, see (3.21), since the necessary rescaling properties of the discrete principal eigenvalue of $\Delta^{\mathrm{d}} + \xi$ and the necessary continuity properties of the map $\varphi \mapsto \lambda^{\mathrm{c}}(Q_R, \varphi)$ are difficult to obtain or to approximate in the most obvious topologies in which one proves the LDP. The topologies in which one can handle some LDP for the rescaled potential are typically much weaker than those in which the eigenvalue is continuous. For this reason, there are not many rigorous proofs in the literature that are based on the methodology described in Sect. 3.2. One important example is the celebrated method of enlargement of obstacles by Sznitman, see his monograph [Szn98] and a very short summary in Sect. 4.7.

4.3 Periodisation

Since the asymptotic methods described in Sect. 4.2 work only for random walks confined to a box having a certain size (possibly depending on t), we first need to find upper and lower bounds for the moments of $U(t)$ in terms of its versions on these boxes. As we explained in Remark 2.1.3, the lower bound is easily obtained, since $U \geq U_B$, where we recall that U_B is the total mass of the solution to the PAM in the box B with zero boundary condition. In Remark 3.1 we described how to control the difference $U - U_B$, but this technique is successful only for very large boxes B, depending on t. This method is meant to introduce just *some* finite horizon to the problem, but not the proper one, whose radius we called $\alpha(t)$. In this section, we rather explain how to derive an upper bound for the moments of U in terms of an *arbitrary* box B, which we afterwards can choose optimal. For this, we have to use periodic boundary condition, which we introduced in Remark 1.6.

Lemma 4.4 (Upper bound via periodisation) *For any box $B = (-R, R]^d \cap \mathbb{Z}^d$, we have the inequality*

$$\langle U(t)\rangle \le \langle U_B^{(\mathrm{per})}(t)\rangle, \qquad t > 0. \tag{4.4}$$

Here is the proof. Denote by $\ell_t^{(R)}(z) = \int_0^t \delta_z(X_s^{(R)})\,\mathrm{d}s$ the local times of the periodised random walk $X^{(R)}$ in the box B, which can be realized by taking the free simple random walk $X = (X_t)_{t\in[0,\infty)}$ modulo B, i.e., $X_t^{(R)} = X_t (\mathrm{mod}\, R\mathbb{Z}^d)$. Then it is not difficult to see that the local times of $X^{(R)}$ are given by

$$\ell_t^{(R)}(z) = \sum_{x\in\mathbb{Z}^d} \ell_t(z + 2Rx), \qquad z \in B, t \in (0,\infty),$$

where ℓ_t is the local time of X. Now using that H is convex and that $H(0) = 0$, we see that H is also super-additive, i.e., $H(l) + H(l') \le H(l + l')$ for any $l, l' \in (0,\infty)$. Indeed, first we use convexity and $H(0) = 0$ to see that $H(\lambda l) = H(\lambda l+(1-\lambda)0) \le \lambda H(l) + (1-\lambda)H(0) = \lambda H(l)$ for any $l \in [0,\infty)$ and $\lambda \in [0,1]$, and then we see that

$$H(l)+H(l') = H\big(\tfrac{l}{l+l'}(l+l')\big)+H\big(\tfrac{l'}{l+l'}(l+l')\big) \le \tfrac{l}{l+l'}H(l+l')+\tfrac{l'}{l+l'}H(l+l') = H(l+l').$$

The super-additivity now shows that the interaction term for the free walk on $\mathbb{Z}^d$ is upper bounded by the same term for the periodised walk:

$$\begin{aligned}\sum_{z\in\mathbb{Z}^d} H(\ell_t(z)) &= \sum_{z\in B}\sum_{x\in\mathbb{Z}^d} H(\ell_t(z + 2Rx)) \\ &\le \sum_{z\in B} H\Big(\sum_{x\in\mathbb{Z}^d} \ell_t(z + 2Rx)\Big) = \sum_{z\in\mathbb{Z}^d} H(\ell_t^{(R)}(z)).\end{aligned} \tag{4.5}$$

Using this in (2.14) (noting the analogous formula for $U_B^{(\mathrm{per})}$), we arrive at the assertion in (4.4).

Lemma 4.4 is indeed one of the canonical starting points for proofs of upper bounds for the moments of $U(t)$, as we explained in Remark 2.1.4. This device is sometimes called periodisation; it is one of a couple of methods to 'compactify' the space. It seems that [DonVar75] is the first work on the PAM that uses this technique.

In order to later derive the proper conclusions, one has to choose the radius of the box B as $R\alpha_t$ (using the notation of the heuristics in Sect. 3.3).

4.4 Spectral Domain Decomposition

Another technique of 'compactification' is demonstrated in [GärKön00, Proposition 1] in the continuous setting and was transferred to the discrete setting in [BisKön01, Prop. 4.4]. This technique works directly for expressions of the form (4.2) and allows for an upper bound for 'global' eigenvalues in terms of 'local' eigenvalues, i.e., of the same expression with a much smaller box instead of B, which might be rather large, as we recall. The error is of the inverse order of the square of the radius of the small box, which is just what one can handle by picking parameters appropriately. The estimate works for the eigenvalue with arbitrary fixed potential, i.e., has nothing to do with randomness and it is not restricted to taking moments. We write $B_r = [-r, r]^d \cap \mathbb{Z}^d$.

Lemma 4.5 (Spectral domain decomposition, discrete version) *There is a constant C that depends only on d such that, for any potential $V: B_{\mathcal{R}} \to \mathbb{R}$,*

$$\lambda^{d}(B_{\mathcal{R}}; V) \le \frac{C}{r^2} + \max_{z \in B_{\mathcal{R}}} \lambda^{d}(z + B_r, V), \qquad \mathcal{R} > r > 0. \tag{4.6}$$

In the proof one sees that already the maximum over much less (carefully chosen) boxes serves as an upper bound, but the above formulation is simpler and is good enough for the application to the PAM. One constructs a partition of the one consisting of smooth functions that approach the indicator functions on the interiors of the subboxes, where we mean 'interior' in the sense that these parts of the subboxes build a decomposition of $B_{\mathcal{R}}$; the overlapping regions are used for pressing the functions down from 1 to 0 in a sufficiently smooth way. The error term comes from the energy of these parts, i.e., from the ℓ^2-norm of their gradients. Reference [BisKön01, Prop. 4.4] is formulated for nonpositive potentials V only, but an inspection of the proof reveals that it actually holds for all potentials.

Here is the continuous version, see [GärKön00, Proposition 1]. For any smooth region $Q \subset \mathbb{R}^d$ and any Hölder-continuous potential $V: Q \to \mathbb{R}$, $\lambda^{c}(Q, V)$ denotes the principal Dirichlet eigenvalue of $\Delta + V$ in Q, and we write Q_r for $[-r, r]^d$.

Lemma 4.6 (Spectral domain decomposition, continuous version) *For any $r \ge 2$, there is a continuous function $\Phi_r: \mathbb{R}^d \to [0, \infty)$, whose support is contained in the one-neighbourhood of the grid $2r\mathbb{Z}^d + \partial Q_r$, such that, for any $\mathcal{R} > r$ and any Hölder-continuous potential $V: \mathbb{R}^d \to \mathbb{R}$,*

$$\lambda^{c}(Q_{\mathcal{R}}, V - \Phi_r) \le \max_{z \in \mathbb{Z}^d : |z| \le \mathcal{R}/r} \lambda^{c}(2rz + B_r, V). \tag{4.7}$$

Moreover, Φ_r can be chosen $2r$-periodic in each coordinate and such that $\int_{Q_r} \Phi_r \le K|Q_r|/r$ for some constant K that does not depend on r.

Let us explain the application in the discrete-space setting. One takes $\mathcal{R}$, depending on t, so large that the a priori bound in (3.4), in conjunction with (3.5), gives a negligible error for $\langle U(t) - U_{B_\mathcal{R}}(t)\rangle$. (Recall that $\mathcal{R}$ is then typically much larger than t.) Hence, the B of (4.2) is the $B_\mathcal{R}$ of Lemma 4.5. The maximum over $z \in B_\mathcal{R}$ is turned into a sum of exponentials, using that they all have the same distribution. Summarizing, we get the following estimate for $r < \mathcal{R}$:

$$\begin{aligned}\left\langle e^{t\lambda^{\mathrm{d}}(B_\mathcal{R},\xi)}\right\rangle &\leq \left\langle e^{t\max_{z\in B_\mathcal{R}}\lambda^{\mathrm{d}}(z+B_r,\xi)}\right\rangle e^{tC/r^2} \leq \sum_{z\in B_\mathcal{R}} \left\langle e^{t\lambda^{\mathrm{d}}(B_r,\xi)}\right\rangle e^{tC/r^2} \\ &= \left\langle e^{t\lambda^{\mathrm{d}}(B_r,\xi)}\right\rangle |B_\mathcal{R}| e^{tC/r^2}.\end{aligned}$$

Hence, the big size of the box is finally tamed down to a big pre-factor, which is negligible with respect to the exponential asymptotics that we are after. Now we choose $r = R\alpha(t)$, with R and $\alpha(t)$ as in the heuristics in Sect. 3.3. This reduces the problem to the appropriate size on which one can use the LDP of Lemma 4.3 for L_t. The error term $C/R^2\alpha(t)^2$ brings the last term onto the scale of the LDP, and it vanishes in the limit $t \to \infty$, followed by $R \to \infty$, two limits that have to be taken finally anyway.

4.5 Cutting

Difficulty (2) for the expression in (4.1) is sometimes handled by some cutting technique, which requires serious work on a case-by-case basis. The problem arises if $\widehat{H}$ is not bounded. The basic idea is to replace $\widehat{H}$ by some cut-off version $\widehat{H}_M$ that is bounded, and controlling the remainder $\widehat{H}^{(>M)} = \widehat{H} - \widehat{H}_M$ with the help of some additional argument in the limit $M \to \infty$, after the limit $t \to \infty$ has been taken. A typical choice is $\widehat{H}_M(l) = \widehat{H}(l) \wedge M$. One separates the factors $e^{\gamma_t \int \widehat{H}_M(L_t)}$ and $e^{\gamma_t \int \widehat{H}^{(>M)}(L_t)}$ from each other using Hölder's inequality with parameters p, q satisfying $\frac{1}{p} + \frac{1}{q} = 1$, such that the first term appears in the p-norm with p very close to one. The exponential rate of the second term is shown to vanish in the limit $t \to \infty$, followed by $M \to \infty$, and then the rate of the first term is shown to approach the desired one in the limit $t \to \infty$, followed by $M \to \infty$ and $p \downarrow 1$.

The details of such an approach must be carefully carried out on a case-by case basis, depending on availabilities of good upper bounds for $\widehat{H}^{(>M)}$ and additional techniques for controlling the rate of the corresponding term. E.g. in [HofKönMör06], some additional elegant inequalities could be employed to arrive in a setting, where $\widehat{H}^{(>M)}$ could be replaced by a negative power of M times a monomial with a small power, and then combinatorial techniques were used to bound the exponential rate in terms of bounds for high polynomial moments.

4.6 Smoothing

Difficulty (3) for the expression in (4.1) is often taken care of by some *smoothing procedure*, i.e., by a replacement of the rescaled local times with the convolution with a smooth approximation of the Dirac measure. This procedure is isolated in technical lemmas in a number of papers, also for the Brownian case.

As an example, let us formulate a version for Brownian motion. Let $\psi: \mathbb{R}^d \to [0,\infty)$ be a rotationally invariant, non-negative, smooth function with support in $[-1,1]^d$ and integral equal to one. For $\delta > 0$, let $\psi_\delta(x) = \delta^{-d}\psi(x/\delta)$. We denote by $\star$ the convolution, that is, $u \star v(x) = \int_{R^d} u(x-y)v(y)\,dy$ for integrable functions $u, v: \mathbb{R}^d \to \mathbb{R}$. The main idea is that, for any integrable u, the function $u \star \psi_\delta$ is smooth and approaches u in the limit $\delta \downarrow 0$ in L^p-sense for any $p \geq 1$ (see [LieLos01], e.g.). Here is a version of this fact that works in the sense of large deviations, see [AssCas03, Lemma 3.1]. By $\mu_t = \frac{1}{t}\int_0^t \delta_{Z_s}\,ds$ we denote the normalised occupation measure for a Brownian motion $(Z_s)_{s\in[0,\infty)}$ in $\mathbb{R}^d$, starting from 0 under $\mathbb{P}_0$.

Lemma 4.7 (Smoothing the occupation measure of Brownian motion) *For any* $\varepsilon > 0$,

$$\lim_{\delta\downarrow 0}\lim_{t\to\infty}\frac{1}{t}\log \sup_{\substack{u:\mathbb{R}^d\to[-1,1]\\ \text{measurable}}} \mathbb{P}_0\big(|\langle \mu_t, u - u\star\psi_\delta\rangle| > \varepsilon\big) = -\infty. \tag{4.8}$$

That is, the probability for the replacement error of the potential u by a smoothed version $u \star \psi_\delta$ being larger than a small amount is enormously small on the exponential scale in $t \to \infty$, if δ is taken small afterwards. Note that the application of Lemma 4.7 assumes a bounded potential, which might require a cutting pre-step; see Sect. 4.5.

For handling Difficulty (3) for the expression in (4.1), i.e., for the rescaled local times L_t of a random walk as defined in (3.28), one needs a version of Lemma 4.7 for this setting. This is provided in [GanKönShi07, Lemma 3.5], which we formulate for the setting of Lemma 4.3.

Lemma 4.8 (Smoothing the rescaled local times of the random walk) *For any* $\varepsilon > 0$,

$$\lim_{\delta\downarrow 0}\lim_{t\to\infty}\frac{\alpha(t)^2}{t}\log \sup_{\substack{u:\mathbb{R}^d\to[-1,1]\\ \text{measurable}}} \mathbb{P}_0\big(|\langle L_t, u - u\star\psi_\delta\rangle| > \varepsilon\big) = -\infty. \tag{4.9}$$

4.7 Method of Enlargement of Obstacles

In the spatially continuous case (i.e., on $\mathbb{R}^d$ instead of $\mathbb{Z}^d$), there is an additional problem to the control of the expression in (4.2): the infinite (even uncountable) combinatorial complexity of the space. In the 1990s, Sznitman contributed a lot to the understanding and the proof techniques for Brownian motion among Poisson obstacles, see his monograph [Szn98]. In particular, he developed proof methods that follow the physical picture, i.e., the interpretation in terms of spectral properties of the random Schrödinger operator. For this, he had to overcome also the problem of making the principal eigenvalue, seen as a function of the random Poisson field, amenable to a large-deviation analysis. For this, he developed a method, a coarse-graining scheme, that yields an upper bound by increasing and discretising the random potential in a careful way. This method was called the *method of enlargement of obstacles*, since the obstacles (the regions in the neighbourhood of the Poisson points) are enlarged by this procedure, giving an upper bound, which is the crucial point. The method is natural, but also involved and introduces three length scales.

Since the method is explained at length in Sznitman's monograph [Szn98] and is briefly surveyed in [Kom98], and since we decided to concentrate on the spatially discrete case in this text, we abstain from trying to explain the method here. A spatially discrete variant of this method was carried out in [Ant94] and [Ant95], but has not been deeper exploited since then, to the best of our knowledge.

4.8 Joint Density of Local Times

A quite sophisticated technique for overcoming the lack of boundedness and continuity for the expression in (4.1) was derived in [BryHofKön07] and makes it possible to derive the LDPs of Lemmas 4.1 and 4.3 even in the *strong* topology, i.e., in the topology that is induced by test integrals against bounded measurable functions (not only bounded continuous functions). This method is based on an explicit upper bound for the joint density of the family of local times for random walks on a finite state space. It was applied in [HofKönMör06, Section 5] to expressions like the one in (4.1). A drawback of this strategy is an error term that makes it fail in too large boxes. The precise formulation of the crucial assertion is as follows (see [BryHofKön07, Theorem 3.6]).

Lemma 4.9 *Let $B \subset \mathbb{Z}^d$ be finite and A_B the generator of a continuous-time Markov chain $(X_t)_{t\in[0,\infty)}$ on B with local times $\ell_t(z) = \int_0^t \mathbb{1}_{\{X_s=z\}}\,\mathrm{d}s$. Then, for* *any measurable function* $F\colon \mathcal{M}_1(B) \to \mathbb{R}$ *and for any* $t > 0$,

$$\mathbb{E}\Big(\exp\Big\{tF\big(\tfrac{1}{t}\ell_t\big)\Big\}\Big) \le \exp\Big\{t \sup_{\mu\in\mathcal{M}_1(B)} \Big[F(\mu) - \| (-A_B)^{1/2} \sqrt{\mu}\|_2^2\Big]\Big\} C_t(B), \tag{4.10}$$

where the error term is given by $C_t(B) = \exp\big\{|B| \log\big(2d\sqrt{8\mathrm{e}}\,t\big)\big\}|B|\mathrm{e}^{|B|/4t}$.

This certainly applies also to simple random walk in a box with any boundary condition. The great value of Lemma 4.9 is that the function F is neither assumed to be continuous nor bounded, and the error term is explicit. The main term on the right-hand side is precisely the variational formula that the large-deviation principle for $\frac{1}{t}\ell_t$ (see Sect. 4.2), in combination with Varadhan's lemma, suggests. After applying some pre-compactification (e.g. by periodisation), the estimate in (4.10) can immediately be applied to the expression in (4.1).

However, to mention also the drawbacks, we also note that the function F that needs to be picked for the rescaled version of L_t in (4.1) depends on t in a non-trivial way, and the appropriate box B as well. As a result, the variational formula on the right-hand side of (4.10) needs to be studied further, and techniques from the theories of Gamma-convergence and finite elements need to be adapted in order to derive a precise asymptotic. In similar situation (intersection local times, i.e., F being a p-norm, respectively the p-th power of the p-norm), this was carried through in [BecKön12] and in [Sch10], respectively.

Lemma 4.9 stands in the tradition of the search for more explicit, deeper and more direct evidence for an interpretation of the family of local times of a continuous-time random walk as a Gaussian process with covariance structure given in terms of the generator of the walk, see the literature remarks in [BryHofKön07].

4.9 Dynkin's Isomorphism

Another fruitful attempt in the search for Gaussian descriptions of the family of local times of random walks is called the *Dynkin isomorphism theorem* [Dyn88]. In a version derived by Eisenbaum it proved extremely useful in the mathematical treatment of upper tails (equivalently, to the high exponential moments) of the self-intersection local times of the walk, which corresponds to the choice $\widehat{H}(l) = l^p$ for some $p > 1$ in (4.1) and taking the p-th root of the functional $\int L_t(x)^p\,\mathrm{d}x$. The Dynkin isomorphism has not yet been applied to the PAM, but is very likely to give also here very good results. Its application to self-intersection local times was initiated in [Cas10] and was brought to full bloom in [CasLauMél14]. We cite here the version by Eisenbaum [Eis95].

Lemma 4.10 (Dynkin's isomorphism) *Let* $X = (X_s)_{s\in[0,\infty)}$ *be a random walk on a finite set* B *with local times* ℓ_t, *and let* τ *be an exponentially distributed random variable, independent of the walk. By* $G = G_{\lambda,B}$ *we denote the Green's function* *of the walk stopped at time* τ. *Let* $(Z_x)_{x\in B}$ *be a centred Gaussian process with covariance matrix G, independent of* τ *and of the walk. For* $s \in \mathbb{R} \setminus \{0\}$, *consider the process* $S_x := \ell_\tau(x) + \frac{1}{2}(Z_x + s)^2$ *with* $x \in B$. *Then, for any measurable and bounded function* $F\colon \mathbb{R}^B \to \mathbb{R}$,

$$\mathbb{E}\Big[F\big((S_x)_{x\in B}\big)\Big] = \mathbb{E}\Big[F\Big(\big(\tfrac{1}{2}(Z_x + s)^2\big)_{x\in B}\Big)\Big(1 + \tfrac{Z_0}{s}\Big)\Big].$$

Hence, essentially the family of local times, taken at an independent exponential time, is in distribution equal to $\frac{1}{2}$ times the square of a Gaussian family with covariance matrix given by the Green's function of the stopped walk. However, there are a number of alterations, due to the addition of parameter s and the density $1 + Z_0/s$. The great value of Lemma 4.10 comes from the fact that Gaussian processes are a lot better behaved as local times and offer a lot of more techniques for their study, like concentration inequalities and explicit calculations. See [CasLauMél14] for these techniques at work and the monograph [MarRos06] for much more on relations between local times and Gaussian processes.

4.10 Discretisation of the Rayleigh-Ritz Formula

Part of Difficulty (3) for estimating (4.2) from above comes from the fact that the eigenvalue $\lambda^{\mathrm{d}}(B, V)$ is a supremum over a quite large set of functions, in particular over an uncountable set. Let us describe one possible way to overcome this in the spatially continuous case. Here, the Rayleigh-Ritz principle for the principal eigenvalue of $\Delta + V$ in a bounded set $Q \subset \mathbb{R}^d$ with regular boundary (recall (2.20) for the discrete-space case) says that

$$\lambda(Q, V) = - \inf_{\phi\in H_0^1(Q)\colon \|\phi\|_2=1} \mathcal{E}_V(\phi), \qquad \text{where } \mathcal{E}_V(\phi) = \|\nabla\phi\|_2^2 - \int V\phi^2, \tag{4.11}$$

for Hölder-continuous potentials V. One natural approach to derive upper bounds for $\lambda(Q, V)$ is to approximate the infimum over this large set by the infimum over a much smaller, actually finite, set of functions that lie so dense that the replacement error is small. One example in the literature where this has been carried out are [MerWüt01a, MerWüt02], from which we cite now. We write Q_R for $[-R, R]^d$.

Lemma 4.11 (Discretisation of the Rayleigh-Ritz formula) *For any* $\eta > 0$, *there are* $M > 0$ *and* $R \geq 1$ *and a finite set* $\Phi_R \subset \{\phi \in \mathcal{C}^1(Q_{R+1})\colon \|\phi\|_2 = 1\}$ *such that, for any* $\mathcal{R} > R$, *and for any Hölder continuous potential* $V\colon Q_{\mathcal{R}} \to \mathbb{R}$,

$$\lambda(Q_{\mathcal{R}}, V) \leq - \min_{y\in Y_{R,\mathcal{R}}} \min_{\phi\in\Phi_R} \mathcal{E}_{V\wedge M}\big(\phi(\cdot - y)\big) + \eta, \tag{4.12}$$

where

$$Y_{R,\mathcal{R}} = \Big\{y \in \frac{R}{\sqrt{d}}\mathbb{Z}^d \colon (y + Q_R) \cap Q_{\mathcal{R}} \neq \emptyset\Big\}.$$

The minima extend over finite sets and can, for applications to the PAM, safely be replaced by sums, to obtain suitable bounds. Indeed, to apply this lemma to (4.2), one estimates the exponential of the eigenvalue (now with V the random potential under consideration) from above by the sum of the exponentials of $-\mathcal{E}_{V\wedge M}(\phi(\cdot - y))$ over y and ϕ. Then one uses the shift-invariance of the distribution of V to get rid of the y-dependence. This turns the problem of estimating (4.2) into the problem of estimating $\langle e^{t(\int (V\wedge M)\phi^2 - \|\nabla\phi\|_2^2)}\rangle$ for a fixed, normalised $\phi \in \mathcal{C}^1(Q_{R+1})$, which is a much easier problem. For the potential V considered in [MerWüt01a, MerWüt02], this problem is handled by using a spatial discretisation technique that is adapted to the Poisson nature of the potential V.

4.11 Compactification by Local-Masses Decomposition

An interesting idea for compactifying the set of probability measures on $\mathbb{R}^d$ was recently introduced in [MukVar15]. The idea is to register only local, widely separated regions in which a non-trivial part of the mass of a given probability measure is located and to forget about their locations. In this way, the measure is represented just by an at most countable collection of sub-probability measures whose total masses add up to not more than one, and whose locations can be thought of as being at the origin. The authors have predominantly in mind to extend the well-known Donsker-Varadhan-Gärtner LDP (see Sect. 4.2.1) for the normalised occupation measures of a Brownian motion from bounded regions to the entire space $\mathbb{R}^d$, but it is clear that the idea can be translated into the $\mathbb{Z}^d$-setting as well. It seems that the method has a great potential to contribute to a deeper investigation of the PAM and possibly also for the one of certain variational formulas like the ones appearing in Sect. 3.4.

Here is a formulation of the compactification idea of [MukVar15]. We intuitively describe how to extract a converging subsequence from a sequence $(\mu_n)_{n\in\mathbb{N}}$ of probability measures on $\mathbb{R}^d$. Consider the concentration function $q_n(r) = \sup_{x\in\mathbb{R}^d} \mu_n(x + [-r, r]^d)$ for any $r \in (0, \infty)$ and extract some subsequence (denoted $(\mu_n)_n$) along which $q_n(r)$ converges towards some $q(r) \in [0, 1]$. By monotonicity, the limit $p_1 = \lim_{r\to\infty} q(r)$ exists in $[0, 1]$. Then there is a sequence of shift vectors $a_n \in \mathbb{R}^d$ such that, along some subsequence, the shifts $\mu_n \star \delta_{a_n}$ of μ_n vaguely converge towards some sub-probability measure α_1 on $\mathbb{R}^d$ with total mass equal to p_1. (We keep writing $\star$ for convolution, so $\mu \star \delta_a$ is the a-shift of μ.) Hence, we can write $\mu_n \star \delta_{a_n} = \alpha_n^{(1)} + \mu_n^{(1)}$ with $\alpha_n^{(1)}$ converging weakly towards α_1 along some subsequence. Now we proceed recursively, i.e., we apply the same procedure to the sequence $(\mu_n^{(1)})_n$ and so forth. In this way, we obtain a sequence

$(\tilde{\alpha}_k, p_k)_{k\in\mathbb{N}}$ of subprobability measures α_k with total masses $p_k \in [0,1]$ satisfying $\sum_{k\in\mathbb{N}} p_k \leq 1$, and we write $\tilde{\alpha} = \{\alpha \star \delta_x : x \in \mathbb{R}^d\}$ for the equivalence class of α with respect to the relation that considers measures as equal if one is a translation of the other. As a result, the measures μ_n concentrate on widely separated compact areas in $\mathbb{R}^d$ of masses p_k, and a mass of $1 - \sum_{k\in\mathbb{N}} p_k$ leaks out. For example, for any $x \in \mathbb{R}^d \setminus \{0\}$, the sequence $(\frac{1}{2}(\delta_{-nx} + \delta_{nx}))_{n\in\mathbb{N}}$ is represented by the sequence $((\frac{1}{2}\tilde{\delta}_0, \frac{1}{2}), (\frac{1}{2}\tilde{\delta}_0, \frac{1}{2}), (0,0), \dots)$, and the sequence of uniform distributions on unboundedly growing balls is represented by the trivial sequence $((0,0), \dots)$, i.e., all the mass leaks out here.

Since the above construction neglects the shift vectors a_n, it actually works in the quotient space $\widetilde{\mathcal{M}}_1(\mathbb{R}^d) = \{\tilde{\mu} : \mu \in \mathcal{M}_1(\mathbb{R}^d)\}$ of $\mathcal{M}_1(\mathbb{R}^d)$. The space $\mathcal{S}$ of all sequences $(\tilde{\alpha}_k, p_k)_{k\in\mathbb{N}}$ of classes of subprobability measures α_k with total masses $p_k \in [0,1]$ satisfying $\sum_{k\in\mathbb{N}} p_k \leq 1$ is the mentioned compactification of $\widetilde{\mathcal{M}}_1(\mathbb{R}^d)$. One can see $\widetilde{\mathcal{M}}_1(\mathbb{R}^d)$ as subset of $\mathcal{S}$ when identifying it with $\{((\tilde{\mu}, 1), (0,0), \dots) \in \mathcal{S} : \mu \in \mathcal{M}_1(\mathbb{R}^d)\}$. On $\mathcal{S}$, [MukVar15] constructs a metric that makes it compact.

It appears that this construction is quite natural and has many benefits and possible extensions. The main application given in [MukVar15] is an LDP for the normalised occupation measures of a Brownian motion (without any restriction to a bounded set), after mapping it into $\widetilde{\mathcal{M}}_1(\mathbb{R}^d)$. The method was used there for proving a tube property for a well-known model from statistical mechanics, the *mean-field polaron model* and was further extended to a deeper study of this model in [KönMuk15] and [BolKönMuk15].

Chapter 5
Almost Sure Asymptotics for the Total Mass

In this section, we explain the basic picture that underlies the almost sure asymptotics of the total mass $U(t)$ under Assumption (H). Like for the moments, we will give an argument only for the lower bound, as this gives a good insight in the behaviour of the PAM, while many proofs of the corresponding upper bounds do not. The heuristics come with a new characteristic variational formula, which is closely connected with the two formulas that describe the moments. Again, we distinguish the spatially discrete case (Sect. 5.1) and the continuous case (Sect. 5.2). It is helpful to recall the considerations made in Sect. 2.3.2 about the difference in the thinking about moments and about almost sure statements.

5.1 Discrete Space

Let us consider the PAM as in (1.1) with initial condition (1.2) with an i.i.d. random potential $\xi = (\xi(z))_{z\in\mathbb{Z}^d}$ having all positive moments $e^{H(t)} = \langle e^{t\xi(0)}\rangle$ finite; in particular, the condition (1.5) holds. As in Sect. 3.2, we suppose that Assumption (H) holds; see (3.26). We want to reveal the mechanism that is behind the almost sure asymptotics of the total mass $U(t)$ of the solution $u(t,\cdot)$.

Recall from (3.3) and (3.6) that $U(t)$, at least as it concerns the moments, can be well approximated with the help of the principal eigenvalue $\lambda^{\mathrm{d}}(B^{(t)},\xi)$ of the Anderson Hamiltonian $\Delta^{\mathrm{d}} + \xi$ in a sufficiently large centred box $B^{(t)}$, i.e., $U(t) \approx e^{t\lambda^{\mathrm{d}}(B^{(t)},\xi)}$, see also Remark 3.1. Hence, it suffices to study the asymptotics of $\lambda^{\mathrm{d}}(B^{(t)},\xi)$ as $t \to \infty$. As we are considering a potential distribution with all positive exponential moments finite, the diameter of $B^{(t)}$ can be taken of order t with logarithmic corrections. For definiteness, we again assume that $\alpha(t) \to \infty$ as $t \to \infty$ (i.e., class (B) or (AB)), but the same heuristics applies in all other cases with appropriate modifications.

W. König, *The Parabolic Anderson Model*, Pathways in Mathematics,
DOI 10.1007/978-3-319-33596-4_5

The idea is to estimate $\lambda^{\rm d}(B^{(t)},\xi) \geq \lambda^{\rm d}(\widetilde{B},\xi)$ for some optimally chosen 'microbox' $\widetilde{B}$ in $B^{(t)}$. (We again use that the eigenvalue is monotonous in the domain.) That is, we search for some local area in which the potential is extremely high and has a particularly good shape, where the latter refers to a particularly large local eigenvalue of $\Delta^{\rm d}+\xi$. This microbox is one of the intermittent islands that we talked about in our phenomenological discussion of intermittency in Sect. 1.4; our lower bound will be based on just one of them.

Recall that we write $B_R = [-R,R]^d \cap \mathbb{Z}^d$ for $R > 0$. Our ansatz is to put $\widetilde{B} = z + B_{R\tilde{\alpha}_t}$ for some $z \in B^{(t)}$ (neglecting a small error close to the boundary of $B^{(t)}$) and for some scale function $\tilde{\alpha}_t \to \infty$ and some radius R (taken large afterwards). Therefore, we consider the shifted and rescaled potential

$$\tilde{\xi}_{t,z}(\cdot) = \tilde{\alpha}_t^2\Big[\xi\big(z+\lfloor\cdot\,\tilde{\alpha}_t\rfloor\big) - h_t\Big] \qquad \text{in } Q_R = [-R,R]^d, \tag{5.1}$$

introducing a new scale h_t for the absolute height of the potential in $\widetilde{B}$. The choice of the pre-factor as the *square* of the spatial scale $\tilde{\alpha}(t)$ is (like in Sect. 3.2) motivated by the asymptotic rescaling properties of the discrete Laplace operator in (3.11).

Scales for the almost-sure behaviour:

$t\log^2 t \approx$ order of diameter of macrobox $B^{(t)}$

$\tilde{\alpha}_t =$ order of diameter of intermittent island

$h_t =$ maximal height of ξ in the macrobox (and in the island)

$\tilde{\alpha}_t^2 =$ reciprocal of the order of deviations of potential from h_t in the island

The choices of $\tilde{\alpha}_t$ and h_t will be determined by (5.3) and (5.6) below.

Note that, for any continuous shape function $\varphi\colon Q_R \to \mathbb{R}$,

$$\tilde{\xi}_{t,z} \approx \varphi \quad \text{in } Q_R \qquad \Longleftrightarrow \qquad \xi(z+\cdot) \approx h_t + \tfrac{1}{\tilde{\alpha}_t^2}\varphi\big(\tfrac{\cdot}{\tilde{\alpha}_t}\big) \quad \text{in } \widetilde{B}-z. \tag{5.2}$$

(Actually, for a lower bound, which we are demonstrating here, only '$\geq$' instead of '$\approx$' is necessary.) For a given z, we want to use the large-deviation principle in (3.14) to derive the probability for the event in (5.2). Indeed, if we could write, for some new scale function $\beta(t) \to \infty$,

$$\tilde{\alpha}_t = \alpha(\beta(t)) \qquad \text{and} \qquad h_t = \frac{H(\beta(t)/\alpha(\beta(t))^d)}{\beta(t)/\alpha(\beta(t))^d}, \tag{5.3}$$

then $\tilde{\xi}_{t,0}$ is identical to $\overline{\xi}_{\beta(t)}$ defined in (3.8). Then an application of (3.14) with $\beta(t)$ instead of t gives that

$$\text{Prob}\big(\tilde{\xi}_{t,0} \approx \varphi \text{ in } Q_R\big) \approx \exp\Big\{-\frac{\beta(t)}{\alpha(\beta(t))^2} I_R(\varphi)\Big\}, \tag{5.4}$$

where we recall the rate function $I_R(\varphi) = \int_{Q_R} J(\varphi(y))\, dy$ from (3.15). Hence, the probability that ξ realizes the event in (5.2) in one of the microboxes (centered at, say, $z = 0$) decays exponentially on the scale $\beta(t)/\alpha(\beta(t))^2$ with rate $I_R(\varphi)$.

Now we examine the probability of the existence of some $z \in B^{(t)}$ such that the event in (5.2) holds for this z. A rough argument is to require that, on an average, there is at least one microbox $z + B_{R\tilde{\alpha}(t)}$ in $B^{(t)}$ in which (5.2) is satisfied, i.e., to put

$$\begin{aligned} 1 \leq \mathbb{E}\Big[\sum_{z\in B^{(t)}} \mathbb{1}\{\tilde{\xi}_{t,z} \approx \varphi \text{ in } Q_R\}\Big] &= |B^{(t)}|\, \text{Prob}\big(\tilde{\xi}_{t,0} \approx \varphi \text{ in } Q_R\big) \\ &\approx t^d \exp\Big\{-\frac{\beta(t)}{\alpha(\beta(t))^2} I_R(\varphi)\Big\}. \end{aligned} \tag{5.5}$$

Here we recall that the volume of the macrobox is roughly t^d with logarithmic corrections, which we neglect. Hence, in order that we can expect one microbox in which the event in (5.2) holds, we should choose $\beta(t)$ according to

$$\frac{\beta(t)}{\alpha(\beta(t))^2} = d \log t, \tag{5.6}$$

that is, $t \mapsto \beta(t)$ is the inverse of the map $s \mapsto s/\alpha(s)^2$, evaluated at $d\log t$. Furthermore, we have to restrict to potential shapes φ that satisfy $I_R(\varphi) \leq 1$. Hence, with the choice of $\beta(t)$ in (5.6) and the choice of $\tilde{\alpha}_t$ and h_t in (5.3), we can expect that at least one $z \in B^{(t)}$ exists such that $\tilde{\xi}_t \approx \varphi$ in Q_R. With all these choices, we obtain the lower bound

$$\frac{1}{t}\log U(t) \approx \lambda^{\mathrm{d}}(B^{(t)}, \xi) \geq \lambda^{\mathrm{d}}\big(\widetilde{B}, h_t + \tfrac{1}{\tilde{\alpha}_t^2}\varphi\big(\tfrac{\cdot}{\tilde{\alpha}_t}\big)\big) \approx h_t + \frac{1}{\tilde{\alpha}_t^2}\lambda^{\mathrm{c}}(Q_R, \varphi), \tag{5.7}$$

where we recall that $\lambda^{\mathrm{c}}(Q, \varphi)$ is the principal Dirichlet eigenvalue of the operator $\Delta + \varphi$ in Q. In the second step of (5.7), we used the shift-invariance of the eigenvalue, and in the last one, we used the rescaling of a 'discrete' eigenvalue to a 'continuous' one, see (3.11) and Remark 3.6. The right-hand side of (5.7) is an optimal lower bound in the limit $R \to \infty$ after optimising on the potential shape φ. In the precision that we follow in these heuristics, this lower bound turns out even to be an asymptotic upper bound, as we will comment on in Remark 5.6. Hence, we can summarise:

Theorem 5.1 (Almost sure asymptotics of the total mass) *Assume that the i.i.d. potential ξ satisfies Assumption (H), and let J be given by* (3.13)*. Then the following holds almost surely.*

(i) If $\alpha(t) \to \infty$ as $t \to \infty$ (equivalently, if $\tilde{\alpha}_t \to \infty$), then

$$\frac{1}{t} \log U(t) = \frac{H(\beta(t)/\alpha(\beta(t))^d)}{\beta(t)/\alpha(\beta(t))^d} - \frac{1}{\alpha(\beta(t))^2}(\tilde{\chi} + o(1)), \qquad t \to \infty, \tag{5.8}$$

where $\beta(t)$ is given in (5.6)*, and*

$$\tilde{\chi} = \inf_{\varphi \in C(\mathbb{R}^d) : \int_{\mathbb{R}^d} J(\varphi(y))\, dy \leq 1} \left[-\lambda^c(\mathbb{R}^d, \varphi) \right]. \tag{5.9}$$

(ii) If $\alpha(t) \to 1$ as $t \to \infty$ (equivalently, if $\tilde{\alpha}_t \to 1$), then $\widehat{H}(y) = \rho y \log y$ for some $\rho \in (0, \infty)$ and

$$\frac{1}{t} \log U(t) = \frac{H(d \log t)}{d \log t} - \tilde{\chi}_\rho + o(1), \tag{5.10}$$

where

$$\tilde{\chi}_\rho = \inf_{\varphi : \mathbb{Z}^d \to \mathbb{R} : \frac{\rho}{e} \sum_{z \in \mathbb{Z}^d} e^{\varphi(z)/\rho} \leq 1} \left[-\lambda^d(\mathbb{Z}^d, \varphi) \right]. \tag{5.11}$$

(iii) if $\alpha(t) \to 0$ as $t \to \infty$ (equivalently, if $\tilde{\alpha}_t \to 0$), then (5.10) *holds with $\rho = \infty$, and $\tilde{\chi}_\infty = 2d$.*

We know from Sect. 3.2 that (i) comprises the two potential classes (B) and (AB) of bounded and almost bounded potentials, while (ii) is the case (DE) of the double-exponential distribution, and (iii) is the case (SP) of the single-peak class. See [BisKön01], respectively Example 5.10, and [HofKönMör06] for precise formulations and proofs of (i) in these two respective cases and [GärMol98] for (ii) and (iii). (We took the freedom to slightly simplify the original theorems; see some of the following remarks.)

Like for the moments in Theorem 3.13, there are two characteristic terms: an H-dependent one that describes the maximal absolute height of the potential in the 'macrobox' $B^{(t)}$, and a second-order term providing some more information about local properties of the potential inside the 'microbox' $\widetilde{B}$, which is one of the intermittent islands that we talked about in Sects. 1.4 and 2.3.

Remark 5.2 (The first term) To achieve a neat and interpretable representation, we actually cheated a bit as it concerns the first term, i.e., the term involving $H(t)$. Actually, this term must be just equal to $\max_{z \in B^{(t)}} \xi(z)$, i.e., the maximum of about

t^d independent copies of $\xi(0)$. Giving deterministic asymptotics for that is standard and relies a bit on the taste and on the way how the assumptions on the upper tails are formulated. In [GärMol98, Theorem 2.2] one finds a different representation of this term, which is derived from the upper tails of $\xi(0)$, but not from the exponential moments. Furthermore, in the class (B) in [BisKön01], the first term is even missing, as it is assumed there that esssup $\xi(0) = 0$, and hence the term $\max_{z\in B^{(t)}} \xi(z)$ is very close to zero, and therefore the first term is negligible with respect to the second. ◇

Note that $\tilde{\alpha}_t$ is logarithmic in t and hence much smaller than $\alpha(t)$, the spatial scale of the intermittent island for the moments. The shape of the potential ξ in the island is determined by the characteristic variational formula; see Remark 5.5. A crucial feature is that the intermittent islands come out of a local optimisation procedure; they are the places with the highest potential values in balls of radius $\tilde{\alpha}_t$, whose size is determined by the crucial condition (5.5).

Remark 5.3 (Technical remark on the lower bound, I) Our heuristics, in particular (5.5), explained only how to find the right scales, but not how to prove the almost sure asymptotics as stated in Theorem 5.1. Many proofs of the lower bound in Theorem 5.1 make crucial use of the Second Borel-Cantelli lemma via the following recipe.

Indeed, one does not consider *all* the possible centerings z of the microbox $\widetilde{B}$, but only certain ones, who have a spacing of $2R\tilde{\alpha}_t$ from each other, i.e., only z in the grid $G = B^{(t)} \cap \lceil 2R\tilde{\alpha}_t \rceil \mathbb{Z}^d$. The benefit is that the boxes $z + B_{R\tilde{\alpha}_t}$ with $z \in G$ are mutually disjoint and therefore the potential values in them are independent, which is a prerequisite for the application of the Second Borel-Cantelli lemma. The cardinality of G is still very close to $|B^{(t)}| \approx t^d$. Now we pick some $\varphi\colon Q_R \to \mathbb{R}$ and derive for the probability of the event in (5.2) with '$\geq$' instead of '$\approx$' the following lower bound.

$$\begin{aligned} \mathrm{Prob}\Big(\exists z \in G\colon \tilde{\xi}_{t,z} \geq \varphi \text{ in } Q_R\Big) &\geq |G| \mathrm{Prob}\big(\tilde{\xi}_{t,0} \geq \varphi \text{ in } Q_R\big) \\ &\approx t^d \exp\Big\{ -\frac{\beta(t)}{\alpha(\beta(t))^2} I_R(\varphi)\Big\} \approx \exp\Big\{ -d \log t \big[I_R(\varphi) - 1\big]\Big\}, \end{aligned} \tag{5.12}$$

where we used (5.4) in the second step and (5.3) in the last one. The first goal is to replace t by some sequence t_n tending to ∞ quickly (e.g., $t_n = 2^n$) and to show that the above lower bound is not summable over $n \in \mathbb{N}$. For doing this, some technicalities have to be overcome, and some slight changes in the objects have to be made, e.g., one should assume that $I_R(\varphi)$ is close to one, and one should replace φ by $\varphi - \varepsilon$ for some small $\varepsilon > 0$, and more. Hence, according to the Second Borel-Cantelli lemma, with probability one, for any sufficiently large n, we know that there is some $z_n \in B^{(t_n)}$ such that something slightly more comfortable than $\tilde{\xi}_{t_n,z_n} \geq \varphi$ in Q_R holds. Afterwards, one has to show that this assertion (or slightly less, but still sufficient) persists for all the values $t \in [t_{n-1}, t_n]$, at least for all sufficiently large $n \in \mathbb{N}$. This shows that, almost surely, for all sufficiently large t, the event in (5.2) holds true for some $z \in B^{(t)}$.

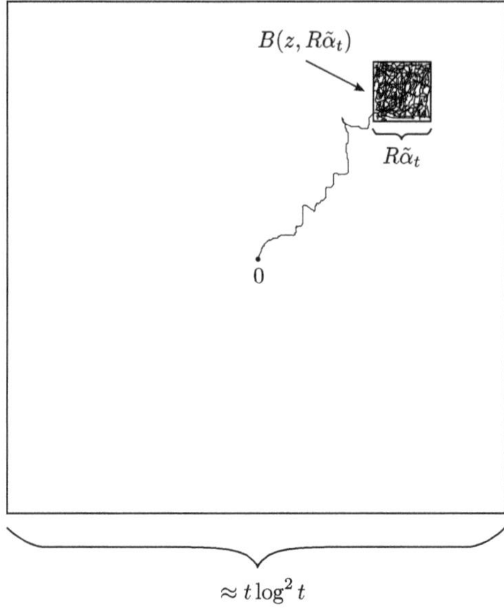

Fig. 5.1 A trajectory that travels quickly to the microbox $\widetilde{B}$ around z with radius $R\widetilde{\alpha}(t)$ and spends the remainder of the time until t in it

The calculation in (5.12) is based on Assumption (J) (see (3.13)), which is usually not the one that is assumed to hold in the literature. Hence, one has to derive (5.4) from the assumption made, which is often Assumption (H). This is explained in Remark 3.11.

Often one does not consider the shape of the rescaled potential, like in the above heuristics, but directly the local principal eigenvalue instead. That is, one derives the existence of a microbox whose principal eigenvalue is close to $\tilde{\chi}_R/\tilde{\alpha}(t)^2$, where $\tilde{\chi}_R$ is the version of $\tilde{\chi}$ in (5.9) that is restricted from $\mathbb{R}^d$ to Q_R. Using this ansatz, the maximisation over the potential shape φ is already built in, but one has to work with precise logarithmic upper tails for the eigenvalue instead of (5.4). These precise upper tails are usually derived via the precise logarithmic asymptotics for the exponential moments of the eigenvalue, which we obtained in the proof of Theorem 3.13. One needs a lower bound for the upper tails, which makes an application of some result in the vicinity of the Gärtner-Ellis theorem necessary, see Remark 3.11. ◇

Remark 5.4 (Technical remark on the lower bound, II) Actually, it is a bit nasty to use the idea of (2.22) for the box $B^{(t)}$ to justify that $\frac{1}{t}\log U(t)$ is lower bounded by the principal eigenvalue of $\Delta^{\mathrm{d}} + \xi$ in $B^{(t)}$ and to proceed with a further restriction to the microbox $\widetilde{B}$. Instead, starting from the Feynman-Kac formula in (2.2), one often uses a lower estimate by inserting the indicator on the event that the random path moves quickly to the box $\widetilde{B}$ (this is done at a negligible cost, if the size of the macrobox $B^{(t)}$ has been chosen suitably) and stays afterwards all the time until t in that box, as is depicted in Fig. 5.1.

The cost for doing the latter is $\exp\{-t\tilde{\alpha}_t^{-2}\lambda^{\rm c}(Q_R,\varphi)\}$ to high precision, which is seen from an application of (2.22) for $B = \widetilde{B}$. ◇

Remark 5.5 (The characteristic variational formula $\widetilde{\chi}$) Of crucial importance for understanding the almost sure asymptotics of the total mass is the *characteristic variational formula* in (5.9) and (5.11), respectively, which describes the shape of the potential in the relevant 'microbox' $\widetilde{B}$, more precisely; the spectral properties of $\Delta^{\rm d} + \xi$ in that microbox. Indeed, only potential shapes $\varphi\colon \mathbb{R}^d \to \mathbb{R}$ satisfying $I(\varphi) \leq 1$ are admitted, where we recall the infinite-volume version of the LDP rate function $I_R(\varphi) = \int_{Q_R} J(\varphi(x))\,{\rm d}x$ in (3.15) with the rate function J introduced in (3.13). The condition $I(\varphi) < 1$ guarantees that the shape φ is not too improbable for a random potential to resemble. Note that, for all examples that we presented in Sect. 3.4, this condition implies that $\lim_{|x|\to\infty} \varphi(x) = -\infty$, i.e., the operator $\Delta + \varphi$ has a compact resolvent, and its infinite-volume principal eigenvalue $\lambda^{\rm c}(\varphi)$ is well-defined, and its Rayleigh-Ritz formula possesses minimisers that are unique, up to spatial shifts. Among all these shapes φ, we optimise this eigenvalue.

Certainly, one expects that the best contribution to the total mass of the PAM comes from those random potential shapes that are particularly close (say, in supremum norm) to the maximiser in (5.9) and (5.11), respectively, but proving this needs much care; see Remark 5.9. ◇

Remark 5.6 (The upper bound in (5.8)) Like for the moment asymptotics in Theorem 3.13, the proof of the upper bound in Theorem 5.1 is technically more involved and more abstract, since one has to take care of *all* the paths in the Feynman-Kac formula respectively *all* the subboxes of $B^{(t)}$ or sizes of all orders, not only some optimal ones. Some of the methods outlined in Sect. 4 are helpful also for the proof of the upper bound in (5.8); we give some guidance here.

Indeed, Lemma 4.5 on the spectral domain decomposition works directly on the principal eigenvalues and is precisely what one needs and gives the desired upper bound for the macrobox eigenvalue in terms of the maximum of the microbox eigenvalues. A Borel-Cantelli argument derives the asymptotics of this maximum, based on the upper tails of one eigenvalue, which one has gained from the asymptotics of the exponential moments of the eigenvalue in the course of the proof of the moment asymptotics in Theorem 3.13. See [BisKön01, HofKönMör06].

Certainly, also variants of Sznitman's method of enlargement of obstacles [Szn98] (see [Ant94]) are suitable to yield a proof in the cases of Brownian motion among Poisson obstacles and survival problems for the simple random walk, respectively.

Furthermore, also the discretisation technique for the Rayleigh-Ritz principle of Lemma 4.11 can be employed for deriving the corresponding upper bound, which has been demonstrated in the spatially continuous case (actually, being reduced to the discrete case in the course of the proof) for a rescaled potential in [MerWüt01a, MerWüt02], see also Sect. 7.3.3.

Both Lemma 4.9 on the joint density of the local times of the random walk and Dynkin's isomorphism in Lemma 4.10 are on the first sight suitable for deriving proofs for almost sure upper bounds for $U_{B(t)}(t)$. However, the relatively huge error term in the estimate of Lemma 4.9 and the complexity of the variational formula on the right-hand side of (4.10), and, respectively, the alterations that one has to make in order to apply Dynkin's theorem let the benefits of each of the two methods appear rather limited.

Note that the periodisation method of Lemma 4.4 is not suitable, since it shovels all the mass of the random walk trajectory into one single box; this type of upper bound works only for moments. ◇

Remark 5.7 (Relation between the variational formulas χ and $\tilde{\chi}$) The variational formulas for $\tilde{\chi}$ in (5.9) and for χ in (3.23) are in close connection to each other. In particular, it can be shown on a case-by-case basis that their minimizers φ differ from each other just by a rescaling (as in class (B); see [BisKön01] and Example 5.10), or by adding a constant (as in class (AB); see [HofKönMör06]), or they are identical (as in class (DE), see [GärMol98]). In the case (SP), they all are trivial, as they are equal to $-\infty$ in all but one site. Recall that the minimiser(s) have the interpretation of the optimal potential shape of the potential in the intermittent islands: $\tilde{\chi}$ for quenched, χ for annealed. ◇

Remark 5.8 (Screening in one dimension) In Remark 5.4, we say that a lower bound is obtained by requiring in the Feynman-Kac formula that the particle runs at high speed to the microbox $\widetilde{B}$, and we say that the cost of this is negligible. In dimensions $d \geq 2$, the negligibility is indeed always true as long as the potential is either $> -\infty$ everywhere or with sufficiently large probability, more precisely, with a probability larger than the critical site-percolation threshold (see [Gri99] for a comprehensive treatment of percolation). In this case, it is known that, almost surely, the set of potential values $\geq -K$, where K is so large that the probability of $\{\xi(0) \geq -K\}$ is also larger than the percolation threshold, contains an infinitely large cluster $\mathcal{C}_K$. As a consequence, there exists, with probability one, for any sufficiently large t, a path from the origin (given that it lies in $\mathcal{C}_K$) to that microbox along which the particle does not lose much mass. Proving this requires two main technical points: (1) one has to make sure that the so-called chemical distance (the number of steps that one needs to walk within $\mathcal{C}_K$) is comparable with the Euclidean distance, and (2) that the microbox can be picked within $\mathcal{C}_K$. See [BisKön01] for details in the case of a bounded potential.

However, in dimension $d = 1$, these arguments do not work anymore, as there is always only precisely one trajectory from the origin to that microbox, and it may happen that the potential assumes too small (i.e., too close to $-\infty$) values such that the particle loses so much mass on the way that the contribution from this sprint is indeed not negligible. This effect is called *screening effect* in [BisKön01a], as the deep valleys screen the mass away from the high peaks. This effect appears as soon as the essential infimum of the potential is equal to $-\infty$ and its lower tails are too thick. See [BisKön01a] for a precise statement. ◇

Remark 5.9 (Almost sure potential confinement) Like for the moments, the above heuristics suggests that the shape of the potential ξ in the peaks, after appropriate shifting and rescaling, resembles the minimising shapes φ^* in the variational formula for $\tilde{\chi}$ in (5.9) (for $\alpha(t) \to \infty$) or in (5.11) (if $\alpha(t) \to 1$), almost surely for large t. Furthermore, one can also conjecture that the solution $u(t, \cdot)$ resembles the corresponding eigenfunction of the operator $\Delta + \varphi^*$ if $\alpha(t) \to \infty$, respectively of $\Delta^{\mathrm{d}} + \varphi^*$ if $\alpha \equiv 1$. This was indeed proved for the case (DE), the double-exponential distribution, in [GärKönMol07], see also Sect. 6.2. ◇

Example 5.10 (Bounded potentials) Let us give explicit formulas for the class (B) of bounded potentials, more precisely, for the single-site distribution given by $\operatorname{esssup} \xi(0) = 0$ and $\log \mathrm{Prob}(\xi(0) > -x) \sim -Dx^{-\gamma/(1-\gamma)}$ for $x \downarrow 0$ with parameters $D > 0$ and $\gamma \in [0, 1)$; see Remark 3.19. We follow [BisKön01]. First we note that $\tilde{\alpha}_t = \alpha(\beta(t)) = (\log t)^{\beta+o(1)}$ and $\beta(t) = (\log t)^{\frac{1}{1-2\nu}+o(1)}$ as $t \to \infty$, where $\nu = \frac{1-\gamma}{d+2-d\gamma} \in [0, \frac{1}{d+2}]$ and $\beta = \frac{2\nu}{1-2\nu} \in (0, \frac{2}{d}]$. Recall from Remark 3.21 that the moment asymptotics are governed by $\hat{\chi}_\circ = \chi_\circ + \rho/(\gamma - 1)$, with $\chi_\circ$ as in (3.52), since the H-terms are dropped in (3.38). Consequently, (5.8) now reads

$$\frac{1}{t} \log U(t) \sim -\frac{1}{\tilde{\alpha}_t^2} \tilde{\hat{\chi}}, \qquad t \to \infty,$$

almost surely, where, instead of (5.9),

$$\tilde{\hat{\chi}} = \inf_{\varphi \in C(\mathbb{R}^d) : \varphi \leq 0, \int |\varphi|^{-\gamma/(1-\gamma)} \leq 1} \left[-\lambda^{\mathrm{c}}(\mathbb{R}^d, \varphi)\right]. \tag{5.13}$$

Note that the map $\varphi \mapsto \int |\varphi|^{-\gamma/(1-\gamma)}$ differs from J, which accounts for dropping the H-terms on the right of (5.8).

Interestingly, the characteristic number $\tilde{\hat{\chi}}$ turns out to be given as

$$\tilde{\hat{\chi}} = \hat{\chi}_\circ^{\frac{1}{1-2\nu}} (1 - 2\nu) \Big(\frac{2\nu}{d}\Big)^{\beta},$$

The main point behind this formula is a rescaling of the form $\varphi_c(x) = c^2 \varphi(cx)$ between the minimiser(s) φ in (5.13) and of the 'dual' formula for $\hat{\chi}_\circ$,

$$\hat{\chi} = \inf_{\varphi \in C(\mathbb{R}^d) : \varphi \leq 0} \Big(\int |\varphi|^{-\gamma/(1-\gamma)} - \lambda^{\mathrm{c}}(\mathbb{R}^d, \varphi) \Big).$$

5.2 Continuous Space

The almost-sure asymptotics of the total mass of the solution to the PAM have been derived also in the continuous case for some important potential distributions, some of which we present here.

Example 5.11 (Brownian motion among Poisson obstacles) Let us explain the heuristics in the simpler (and historically important) situation of a Brownian motion among Poisson obstacles, see Remark 1.15 and Sect. 3.5.1. For a proof and more details, we refer to [Szn98, Theorem 4.5.1].

As in Sect. 3.5.1, we consider the potential $V(\cdot) = -\sum_i W(\cdot - x_i)$ with a standard Poisson point process $(x_i)_{i\in\mathbb{N}}$ with parameter ν and the cloud $W = \infty \mathbb{1}_{K_a}$, where K_a is the centred ball of radius $a > 0$. Recall from Example 2.10 or from the Feynman-Kac formula in (3.55) that the total mass

$$U(t) = \mathbb{P}_0(Z_s \notin \mathcal{O} \text{ for every } s \in [0,t])$$

is equal to the survival probability, i.e., the probability does the Brownian motion Z does not fall into the union $\mathcal{O} = \bigcup_i K_a(x_i)$ of a-balls around the Poisson points by time t.

We describe how to get an optimal lower bound for $U(t)$. The first step is to restrict to a large t-dependent centred box $B^{(t)}$ whose radius is roughly equal to t (with logarithmic corrections). Inside $B^{(t)}$, we need to find an obstacle-free set A, i.e., a (preferably, large) set $A \subset \mathcal{O}^{\mathrm{c}}$. For the sake of simplicity, we proceed with the simpler requirement that $\omega(A) = 0$, where $\omega = \sum_i \delta_{x_i}$ is the Poisson process. The probability that A contains no Poisson points is equal to $\mathrm{e}^{-\nu|A|}$. If we require the Lebesgue measure $|A|$ so large that this is roughly equal to $1/|B^{(t)}| \approx t^{-d}$, then we may expect that, almost surely, for any large t, there is at least one (random and t-dependent) shift A_t of A in $B^{(t)}$ that contains no Poisson points. (The reason is that there are about t^d such shifts, i.e., t^d independent trials for obstacle-free places; a Borel-Cantelli argument gives the statement.) Hence, we need to consider only sets A of Lebesgue measure of size $\frac{d}{\nu}\log t$.

Now we obtain a lower bound for $U(t)$ by restricting to the event that the Brownian motion travels to A_t within some small time interval of length $o(t)$ and then stays in A_t for the remaining time interval of length $t - o(t)$. Similarly to the explanations in Remark 5.8, one shows that the contribution that comes from the sprint to A_t is negligible on the first-order scale. This means that the contribution comes dominantly from the long stay in A_t. Recall the large-deviation principle for the normalized occupation measures of the Brownian motion from Sect. 4.2.1 (or from the handy formula (3.56)) and recall from there that the large-time exponential rate of non-exit probabilities from compact sets is equal to the principal eigenvalue. Hence, the contribution from the long stay in A_t is given in terms of the principal eigenvalue $\lambda_1(A_t)$ of Δ in A_t with zero boundary condition, i.e., we obtain the lower bound

$$U(t) \geq \mathrm{e}^{t(\lambda_1(A_t)+o(1))}, \qquad t \to \infty.$$

Certainly, $\lambda_1(A_t) = \lambda_1(A)$, since A_t is just a shift of A. Again, the Faber-Krahn inequality (see [Ban80]) implies that the optimal (i.e., with largest eigenvalue $\lambda_1(A)$) shape of A with given volume is a ball. Since $|A| = \frac{d}{\nu}\log t$, its radius is equal to $\tilde{r} = (\frac{d}{\omega_d \nu}\log t)^{1/d}$, where ω_d is the volume of the unit ball. Hence,

$$\begin{aligned}\frac{1}{t}\log U(t) &\geq (1+o(1)) \sup_{A:|A|=\frac{d}{\nu}\log t} \lambda_1(A) = (1+o(1))\lambda_1(K_{\tilde{r}}) = (1+o(1))\frac{\lambda_1(K_1)}{\tilde{r}^2} \\ &= -(\log t)^{-2/d}(\tilde{c}(d,\nu)+o(1)),\end{aligned} \tag{5.14}$$

where $\tilde{c}(d,\nu) = \lambda_1(K_1)(\omega_d\nu/d)^{2/d}$. This formula is indeed just a special case of (5.9). The complementary bound of (5.14) holds as well. We see from the above that the spatial scale of the radius of the intermittent islands is $(\log t)^{1/d}$, and the islands are perfect balls. Formula (5.14) with '=' instead of '$\geq$' is analogous to (5.8) in the case (B), when the bounded potential has been shifted such that its essential supremum is equal to zero and hence the first term negligible. ◇

Remark 5.12 (Gibbsian point fields) Recall from Remark 3.22 (see also Example 1.17) that in [Szn93] the point process $(x_i)_i$ was assumed to be a Gibbsian point process with a number of assumptions on the pair-interaction functional involved. Under these assumptions, the result for the almost sure behaviour of the total mass is literally identical with the above result; however the proofs are more involved. ◇

Example 5.13 (Gaussian potential) One of the most natural choices of a random potential in $\mathbb{R}^d$ is a continuous Gaussian potential, see Sect. 3.5.2. The almost sure asymptotics of the total mass for a Hölder-continuous centred stationary Gaussian potential $V = (V(x))_{x\in\mathbb{R}^d}$ are identified in [GärKönMol00]. This potential is treated there as one example of a correlated potential satisfying some rescaled large-deviation principle; see Example 7.4. Furthermore, it is also an example of a potential having long-reaching correlations; see Example 7.5.

The main assumptions on the Gaussian potential are as in [GärKön00]. Indeed, the covariance function B of the Gaussian potential V is assumed twice continuously differentiable, hence V can be assumed Hölder continuous with any parameter in $(0,1)$ (see [Adl90] for the theory of regularity of Gaussian fields). Let $L(h) = \sup_{t>0}(ht - H(t))$ denote the Legendre transform of the logarithm of the moment generating function $H(t) = \log\langle e^{tV(0)}\rangle = \frac{1}{2}t^2\sigma^2$ with $\sigma^2 = B(0)$, and define h_t as a solution to the equation $L(h_t) = d\log t$. Then the main result of [GärKönMol00] is

$$\frac{1}{t}\log U(t) = h_t - (\chi + o(1))\sqrt{h_t}, \qquad t\to\infty, \text{ almost surely,} \tag{5.15}$$

where again $\chi = (2\sigma^2)^{-1/2}\mathrm{tr}(\Sigma)$, and Σ^2 is the Hessian matrix of B at zero. Note the formal similarity to the moment asymptotics in (3.62). It is not difficult to prove that

$$h_t = (2d\sigma^2\log t)^{1/2} \sim \max_{x\in[-t,t]^d} V(x), \qquad t\to\infty;$$

it formally coincides with h_t in the heuristics in Sect. 5.1. Again, the first term in (5.15) was earlier derived in [CarMol95]. The second term reflects the heuristics that the main contribution to $U(t)$ comes from a microbox in $[-t,t]^d$ with radius of order $h_t^{-1/4}$, where the potential V approaches the non-random parabolic shape $h_t p$, where $p(x) = 1 - \frac{1}{2\sigma^2}|\Sigma x|^2$, centred at the random localisation centre. The principal eigenvalue of $\Delta + h_t p$ is easily calculated to be $h_t - \chi\sqrt{h_t}$, which is the right-hand side in (5.15). Let us remark that (5.15) was derived under very mild conditions on the decay of $B(x)$ for $x \to \infty$.

Interestingly, the peaks in the Gaussian potential have a parabolic shape, the description of which depends only on $B(0)$ and $B''(0)$, but not on $B^{(4)}(0)$. Indeed, one easily calculates that, for any site $x_0 \in \mathbb{R}^d$, the variables $V(x_0)$, $V'(x_0)$ and $v = V''(x_0) - B''(0)V(x_0)/\sigma^2$ are independent Gaussians. In particular, $V(x_0)$ and $V''(x_0)$ are highly correlated, and large values of $V(x_0)$ enforce large values of $-V''(x_0)$. More precisely, given that V has a large local maximum $V(x_0) \approx h_t$ at x_0, then $V'(x_0) = 0$ and $|v| = |V''(x_0) - B''(0)V(x_0)/\sigma^2| \ll h_t$ and therefore, in a neighbourhood of x_0,

$$V(x) \approx V(x_0) + \frac{1}{2}V''(x_0)(x-x_0)^2 \approx h_t + \frac{1}{2}\Big(\frac{B''(0)}{\sigma^2}V(x_0) + v\Big)(x-x_0)^2$$
$$\approx h - tp(x-x_0).$$

◇

Remark 5.14 (Generalised Gaussian potential) A Gaussian potential V on $\mathbb{R}^d$ with much less regularity was considered in [Che14]. Indeed, following a question raised in [CarMol95], it was assumed that the covariance function B has a behaviour of the form $B(x) \sim \sigma^2|x|^{-\alpha}$ for $x \to 0$ with some $\sigma^2 \in (0,\infty)$ and some $\alpha \in (0,\min\{2,d\})$ (still keeping that the potential V is centred and stationary). Away from the origin, B is assumed to be continuous and bounded. Such a Gaussian potential does not exist as a function, but has to be constructed with the help of smooth test functions; examples are white noise and fractional white noise. The main result of [Che14] is that, almost surely, as $t \to \infty$,

$$\frac{1}{t}\log U(t) \sim (2d\sigma^2\log t)^{\frac{2}{4-\alpha}}$$
$$\times \sup_{g\in H^1(\mathbb{R}^d):\|g\|_2=1}\Big(\Big(\int_{\mathbb{R}^d}\int_{\mathbb{R}^d}\mathrm{d}x\mathrm{d}y\,\frac{g^2(x)g^2(y)}{|x-y|^\alpha}\Big)^{1/2} - \|\nabla g\|_2^2\Big).$$

Note that the right-hand side for $\alpha = 0$ is equal to the first term in (5.15). For $\alpha > 0$, interestingly, the behaviour of the Gaussian potential close the local maxima appears to contribute already to the first term. ◇

Example 5.15 (Poisson shot-noise potential) Another natural choice of the potential is a Poisson shot-noise potential, see Sect. 3.5.3. For such a potential, the almost-sure asymptotics of the total mass of the solution to the PAM have been derived in [GärKönMol00]. Recall from Sect. 3.5.3 the standard Poisson point process $(x_i)_{i\in\mathbb{N}}$ on $\mathbb{R}^d$ with intensity $\nu \in (0,\infty)$ and the non-negative cloud $\varphi\colon \mathbb{R}^d \to [0,\infty)$. Like in Example 5.13, we neglect issues about the decay of the cloud at ∞ and simply assume φ to be compactly supported; see Sect. 7.5 for questions about the correlation length of this potential. We consider the potential $V(x) = \sum_{i\in\mathbb{N}} \varphi(x - x_i)$, i.e., with the opposite sign as in the obstacle case of Sect. 5.11. A mild assumption on φ implies that V is Hölder-continuous. We assume that φ is strictly maximal in 0 with a strictly positive definite Hessian matrix $\Sigma^2 = -\varphi''(0)$. Clearly, $H(t) = \log\langle \mathrm{e}^{tV(0)}\rangle = \nu \int (\mathrm{e}^{t\varphi(x)} - 1)\,\mathrm{d}x$.

Like in the Gaussian case of Example 5.13, we introduce the Legendre transform $L(h) = \sup_{\rho>0}(\rho h - H(\rho))$ of H and define h_t via $L(h_t) = d\log t$. Then it is derived in [GärKönMol00] that the asymptotics in (5.15) literally hold true with the same value $\chi = (2\sigma^2)^{-1/2}\mathrm{tr}(\Sigma)$, where now $\varphi(0) = \sigma^2$, and Σ^2 is the Hessian of $-\varphi$ at zero. The interpretation of this result is the same as in the Gaussian case; we do not spell it out. Again, $h_t \sim \max_{[-t,t]^d} V$, but here the numerical value is $h_t \sim d\sigma^2 \log t/\log\log t$. This asymptotics is too rough to replace h_t in (5.15), as the error term of the first term would spoil the precision of the second. Sufficiently precise asymptotics for h_t would have to depend on more details of the cloud in a neighbourhood of zero. ◇

Chapter 6
Details About Intermittency

In this chapter, we discuss much stronger and more detailed assertions about intermittency than we did before. They characterise the long-time behaviour of the mass flow through a random potential in a quite satisfactory way and give much additional information on the solution to the PAM. In Sect. 6.1 we give an account on these assertions and a survey on the literature and on the remaining sections of this chapter.

6.1 Survey

One of the fundamental properties of the PAM is intermittency. Here we want to understand this phenomenon in a most descriptive way, i.e., in the sense that the mass of the heat flow through the random potential is concentrated in a few small islands, the intermittent islands, which are time-dependent and randomly located. This is a strong effect of *concentration* or *localisation*. We discussed intermittency already on various levels of deepness and from different angles, see Sects. 1.4, 2.2.4 and 2.3. A rough understanding of this effect was utilized in the proof of the lower bound for the almost sure behaviour of the total mass (see Sect. 5.1), where we used the contribution of just one such island to obtain a sharp lower bound, at least on the exponential scale that we looked at. We also analysed the characteristic variational formulas χ and $\chi_\circ$ in Sect. 3.4 and argued that they describe the potential and the solution in the intermittent islands.

W. König, *The Parabolic Anderson Model*, Pathways in Mathematics,
DOI 10.1007/978-3-319-33596-4_6

In this chapter, we want to go much deeper and discuss much more detailed aspects of intermittency:

Deterministic shapes: Inside these islands, the potential ξ and the solution $u(t,\cdot)$ approximate the shapes given by characteristic variational formulas.
Random locations: The locations of the islands form a Poisson point process in space.
Mass concentration: The contribution from the complement of the union of these islands is negligible.
One-city theorem: The mass concentration takes place in just one of these islands.
Ageing: The time-evolution of the entire mass flow shows an ageing behaviour in terms of the city location process.

All these assertions are presumably true in great generality for the solution to the PAM defined in (1.1), (1.2) and have been proved yet for various interesting potential distributions, as we will report on. The first four of these statements concern only one fixed, late time, a snapshot of the mass flow, but the fifth one considers the entire time-evolution of the mass flow.

In the annealed setting, the above aspects are not too interesting, as most of them have been implicitly already convincingly been handled in the treatment of the moments in Sect. 3.2; see also Remarks 3.8 and 3.14 for the shape property. An exception is the mass concentration property, which has been handled for the class (DE) only, to the best of our knowledge, see Sect. 7.1. However, in the quenched setting and in the setting of convergence in probability, they are highly interesting, as they show a rich picture, give rise to deeper study, and their proofs require combinations of quite subtle means.

As usual, we consider the solution $u(t,\cdot)$ of the PAM in (1.1) with localised initial condition $u(0,\cdot)=\delta_0(\cdot)$. The main goal is to explain how to find relatively small and few random subsets $A_1,\dots,A_{n_t}$ of $\mathbb{Z}^d$, the intermittent islands, such that

$$U(t)\sim\sum_{i=1}^{n_t}\sum_{z\in A_i}u(t,z),\qquad t\to\infty,\tag{6.1}$$

i.e., the contribution from outside the union of these islands is negligible w.r.t. the contribution from inside. (Note that this is a much stronger assertion than saying that just one of them gives a comparable contribution.) As we discussed in Sect. 2.2.3, crucial for this is the eigenvalue expansion

$$u_B(t,\cdot)=\sum_{k=1}^{|B|}\mathrm{e}^{t\lambda_k(B)}v_k(0)v_k(\cdot),\qquad t\in(0,\infty),\tag{6.2}$$

(recall (2.18) in Sect. 2.2.1) for the restriction u_B of the solution to some box $B \subset \mathbb{Z}^d$, which we will have to take t-dependent and sufficiently large. From this, we (rightfully) expect that the sets A_i will turn out to be the sets on which the leading eigenfunctions v_k of the Anderson operator $\Delta^{\rm d} + \xi$ are concentrated. However, for a higher precision (i.e., for obtaining (6.1) with a small n_t), also their distance to the origin has to be taken into account, as we will see.

Let us give some survey on the literature and on the remainder of this chapter. Results of the type in (6.1) in the almost-sure sense were first derived in the cases of Brownian motion among Poisson obstacles [Szn98] (see Remark 6.3) and for the double-exponential distribution [GärKönMol07], both with possibly a quite large value of n_t, that is, $n_t = t^{o(1)}$. The latter work derived also the shapes of the potential ξ and of the solution $u(t, \cdot)$ in the peaks, which will be presented in Sect. 6.2. The strongest concentration property that can be hoped for in the almost-sure sense is with $n_t = 2$ for reasons that we explain in Remark 6.7. This was proved for the first time in [KönLacMörSid09], and that for the heaviest-tailed distribution, the Pareto distribution.

In the sense of convergence in probability, concentration with $n_t = 1$ was proved for some related potential distributions in [LacMör12, SidTwa14, FioMui14] (even in random environment, see [MuiPym14] and Sect. 7.9.2). All these cases belong to the (vicinity of the) class (SP) (see Remark 3.16), i.e., the set A_1 in (6.1) is a singleton here. Recently, the class (DE) (see Remark 3.17) was handled in [BisKönSan16], which we outline in Sect. 6.4; here the geometry of the islands is non-trivial.

An important and serious pre-step for the proof of such strong concentration properties is the derivation of a Poisson point process convergence for the point process consisting of all the top eigenvalues and corresponding eigenfunctions in large boxes. Some phenomenological relations with Anderson localisation come to the surface here. Assertions of this type have been derived in a series of papers by Astrauskas, see his recent survey [Ast16], but only for potential distributions in the class (SP), where the eigenfunctions are delta-like functions. In the much more challenging case of the double-exponential distribution, [BisKön16] derived such statements; this is discussed in Sect. 6.3.

Ageing properties were first derived for the Pareto distribution in [MörOrtSid11] and more recently for other potentials in the class (SP) in [SidTwa14, FioMui14]and for the class (DE) in [BisKönSan16]; we give an account on this in Sect. 6.5. Let us mention that [Mör11] is a rather readable survey on the concentration and ageing properties of the PAM with heavy-tailed distributions, in particular the Pareto distribution.

However, for the least heavy-tailed potential classes, (AB) and (B), all the above five detailed assertions have not yet been formulated nor proved in the literature, to the best of our knowledge, with some kind of exception of (B) in the spatially continuous case, i.e., Brownian motion among Poisson obstacles, see [Szn98]. This is a certain lack in the study of the PAM, but it is expected that the overwhelming part of the formulation and of the proofs will be rather similar to the existing proofs in the other cases. Hence, after 25 years of intense research on the PAM, it appears that its standard version for i.i.d. random potentials has, phenomenologically and mathematically, been fully understood, as only some partial questions remain open yet, most of which can be considered of technical nature.

6.2 Geometric Characterisation of Intermittency

In this section we give a more precise formulation of (6.1) in the almost-sure sense for the class (DE) introduced in Remark 3.17. In particular, we describe the sets A_i and the typical shape of the potential ξ and of the solution $u(t,\cdot)$ inside these sets. We follow [GärKönMol07], but slightly simplify some facts.

We assume that the parameter ρ appearing in (3.44) is so large that, up to spatial shifts, the variational problem in (3.47) possesses a unique minimiser, which has a unique maximum [GärHol99]. (Recall that $\rho > 16$ suffices.) By V_* we denote the unique centred maximizer of (3.47), i.e., of the variational formula

$$\chi = \inf_{\varphi\in\mathbb{R}^{\mathbb{Z}^d}} \Big(\frac{\rho}{\mathrm{e}}\sum_{z\in\mathbb{Z}^d} \mathrm{e}^{\varphi(z)/\rho} - \lambda(\varphi)\Big),$$

where we recall that $\lambda(\varphi)$ is the principal eigenvalue of $\Delta^{\mathrm{d}}+\varphi$ in $\mathbb{Z}^d$, and we may restrict the infimum to those φ that satisfy $\lim_{z\to\infty}\varphi(z)=-\infty$. For definiteness, we assume that V^* attains its unique maximum at the origin and call V_* *optimal potential shape*. Some crucial properties of the formula (3.47) are as follows. The operator $\Delta^{\mathrm{d}}+V_*$ has a unique non-negative eigenfunction $w_*\in\ell^2(\mathbb{Z}^d)$ with $w_*(0)=1$ corresponding to the eigenvalue $\lambda(V_*)$. Moreover, $w_*\in\ell^1(\mathbb{Z}^d)$ is positive everywhere on $\mathbb{Z}^d$.

One crucial object is the maximum $h_t=\max_{z\in B^{(t)}}\xi(z)$ of the potential in the centred macrobox $B^{(t)}$, which has radius $\approx t\log^2 t$, as we recall. We shall see that the main contribution to the total mass $U(t)$ comes from the neighbourhoods of the local sets of best local coincidences of $\xi - h_t$ with spatial shifts of V_*. These neighbourhoods are widely separated from each other and hence not numerous. We may restrict ourselves further to those neighbourhoods in which, in addition, $u(t,\cdot)$, properly normalized, is close to w_*.

Denote by $B_R(y)=y+B_R$ the closed box of radius R centered at $y\in\mathbb{Z}^d$ and write $B_R(A)=\bigcup_{y\in A}B_R(y)$ for the R-box neighbourhood of a set $A\subset\mathbb{Z}^d$. In particular, $B_0(A)=A$.

For any $\varepsilon > 0$, let $r(\varepsilon, \rho)$ denote the smallest $r \in \mathbb{N}_0$ such that

$$\|w_*\|_2^2 \sum_{x \in \mathbb{Z}^d \setminus B_r} w_*(x) < \varepsilon. \tag{6.3}$$

Given $f: \mathbb{Z}^d \to \mathbb{R}$ and $R > 0$, let $\|f\|_R = \sup_{x \in B_R} |f(x)|$. The main result of [GärKönMol07] is the following.

Theorem 6.1 *There exists a random time-dependent subset* $\Gamma^* = \Gamma^*_{t \log^2 t}$ *of* $B_{t \log^2 t}$ *such that, almost surely,*

$$(i)\quad \liminf_{t \to \infty} \frac{1}{U(t)} \sum_{x \in B_{r(\varepsilon,\rho)}(\Gamma^*)} u(t,x) \geq 1 - \varepsilon, \qquad \varepsilon \in (0,1); \tag{6.4}$$

$$(ii)\quad |\Gamma^*| \leq t^{o(1)} \quad \textit{and} \quad \min_{y, \tilde{y} \in \Gamma^* : y \neq \tilde{y}} |y - \tilde{y}| \geq t^{1-o(1)} \qquad \textit{as } t \to \infty; \tag{6.5}$$

$$(iii)\quad \lim_{t \to \infty} \max_{y \in \Gamma^*} \left\| \xi(y + \cdot) - h_t - V_*(\cdot) \right\|_R = 0, \qquad R > 0; \tag{6.6}$$

$$(iv)\quad \lim_{t \to \infty} \max_{y \in \Gamma^*} \left\| \frac{u(t, y + \cdot)}{u(t,y)} - w_*(\cdot) \right\|_R = 0, \qquad R > 0. \tag{6.7}$$

Theorem 6.1 states that, up to an arbitrarily small relative error ε, the islands with centers in Γ^* and radius $r(\varepsilon, \rho)$ carry the whole mass of the solution $u(t, \cdot)$. In terms of (6.1), $n_t = |\Gamma^*| = t^{o(1)}$, and the A_i are the R-neighbourhoods of the sites in Γ^*. In each set A_i, the shapes of the potential and the normalized solution resemble $h_t + V_*$ and w_*, respectively. The mutual distances of any two of these island increase almost like t.

One crucial input in the proof of the shape properties (i.e., (iii) and (iv)) is a property of the characteristic variational formula χ in (3.47) that is sometimes called *stability*: If some sequence of admissible functions $\varphi_n: \mathbb{Z}^d \to \mathbb{R}$ is picked such that their values of the functional $\frac{\rho}{e} \sum_z e^{\varphi(z)/\rho} - \lambda(\varphi)$ converges towards the infimum, χ, then, up to some spatial shift, φ_n converges towards the minimiser V^*.

The main strategy of [GärKönMol07] is not based on the eigenvalue expansion in Remark 2.2.1, since it is difficult to handle the possible negativity of the eigenfunctions at zero. Instead, a strategy is developed that works exclusively with *principal* eigenfunctions of $\Delta^d + \xi$ in local neighbourhoods of high exceedances of the potential, after destroying the quality of the eigenvalues in all the other islands. One crucial point is the proof of the exponential localisation of the corresponding eigenfunctions using a decomposition technique for the paths in the Feynman-Kac representation for these principal eigenfunctions (called *probabilistic cluster expansion* in [GärKönMol07]). This is based on the fact that, with high probability, the small regions in which the potential ξ is extremely high, are of bounded size and have a huge distance to each other. Hence, if the path in the Feynman-Kac formula visits more than one of them, then he has to travel for long time through space with much lower potential values, and this implies that their contribution to the expectation is negligible.

Remark 6.2 (Deriving concentration of $u(t,\cdot)$ from concentration of eigenfunctions) Let us present one crucial technical input in the proof of Theorem 6.1, which was used also in later papers on this subject. It shows how to derive the concentration property of the solution $u(t,\cdot)$ of the PAM from the exponential localisation property of the leading eigenfunctions.

We consider the function w (depending on $B = B^{(t)}$ and Γ^*) given by the Feynman-Kac formula

$$w(t,x) = \mathbb{E}_x\Big[\exp\Big\{\int_0^t \xi(X_s)\,\mathrm{d}s\Big\}\delta_0(X_t)\mathbb{1}\{\tau_{B^c} > t\}\mathbb{1}\{\tau_{\Gamma_*} \le t\}\Big], \qquad t \ge 0,\, x \in B, \tag{6.8}$$

where we write $\tau_A = \inf\{t > 0 \colon X_t \in A\}$ for the entrance time into a set $A \subset \mathbb{Z}^d$. Note that w is the Feynman-Kac formula for the solution of the PAM in the set B with zero boundary condition, restricted to those paths that visit the set Γ^*. (Those who don't can be handled by rougher means.) Then [GärKönMol07, Theorem 4.1] says that, for any $t > 0$,

$$w(t,\cdot) \le \sum_{y \in \Gamma^*} w(t,y)\|v_y\|_2^2 v_y(\cdot), \tag{6.9}$$

where v_y is the principal eigenfunction of $\Delta^{\mathrm{d}} + \xi$ in B after removing from each island the site with the highest potential value, with exception of that one that contains y, normalised by $v_y(y) = 1$. (We mentioned these principal eigenfunctions as crucial objects in the proof of Theorem 6.1.) In particular, for any $r \ge 0$ and $t > 0$,

$$\frac{\sum_{x \in B\setminus B_r(\Gamma^*)} w(t,x)}{\sum_{x\in B} w(t,x)} \le \max_{y\in\Gamma^*}\Big[\|v_y\|_2^2 \sum_{x\in B\setminus B_r(\Gamma^*)} v_y(x)\Big], \tag{6.10}$$

which is obviously enormously helpful for proving the concentration property of w in the neighbourhoods of the set Γ^*. The proof of (6.9) consists of clever, but elementary manipulations using the Markov property and related tools. ◇

There is no control on the differences between any two of the top eigenvalues, but it is shown that the concentration centres of these eigenfunctions have mutual distance $t^{1-o(1)}$ from each other. This in turn implies that there are not more than $t^{o(1)}$ of them, and therefore there must be, somewhere close to the top of the spectrum, some gap of minimal size $t^{-o(1)}$. This gap played a crucial rôle in the proof of the exponential localisation.

Remark 6.3 (Brownian path concentration among Poisson obstacles) In the case of Brownian motion among Poisson obstacles (see Remark 1.10) an assertion was proved (see [Szn98]) that is closely related to Theorem 6.1. In fact, this assertion is formulated in terms of the behaviour of the motion among the obstacles rather than in terms of mass concentration of its occupation probabilities. The main result here can be roughly formulated as follows. Almost surely, as $t \to \infty$, there are $n_t = t^{o(1)}$ balls $A_1, \dots, A_{n_t} \subset \mathbb{R}^d$ of radius $\asymp \tilde{\alpha}_t$ with mutual distance $t^{1-o(1)}$ such that the Brownian motion of the Feynman-Kac formula does the following with probability tending to one under the transformed path measure $Q_{\xi,t}$ defined in (2.15). The motion arrives, after some deterministic diverging time $\ll t$, at one of these balls and does not leave it anymore up to time t.

These balls are characterised in terms of the property that they optimise, within the macrobox $B_{t\log^2 t}$, the sum of the principal eigenvalue of $\Delta + V$ in that region and a certain quantity that measures the exponential cost for a Brownian motion to travel to that region through the random potential. The latter can be formally written as a *Lyapounov exponent*, but there exists no explicit formula for it; its existence is based on subadditivity. ◇

Both results in [GärKönMol07] and [Szn98] do not provide much control on the location of the concentration centres of the leading eigenfunctions, i.e., of the sets A_i in (6.1), nor on the minimal number n_t of the relevant islands that are needed to prove (6.1) (only $n_t = t^{o(1)}$ was proved). The reasons were that both did not derive a precise order-statistics statement on the eigenvalues that includes control on the gaps between them, and that both did not include the analysis of the distance of the islands to the origin, nor the size of the term $v_k(0)$ in (6.1) in their investigation.

6.3 Eigenvalue Order Statistics

On the way to a proof of the concentration property in just one island, we have to get sufficient control on the gaps between all the eigenvalues $\lambda_k(B)$ in (6.2) and on their influences via the term $v_k(0)$. More precisely, one has to get control on the largest value of $e^{t\lambda_k} v_k(0)$, $k \in \mathbb{N}$, and to show that the gap to the second-largest of them is significantly smaller. Note that a priori the maximal k does not have to be $k = 1$, i.e., the answer is not necessarily given by the largest eigenvalue. Let us first explain the general mechanism that comes to the surface here; the underlying mathematics is a mixture of Anderson localisation theory from mathematical physics and spatial extreme-value theory from statistics.

6.3.1 Point Process Convergence

In this section, we give a heuristic explanation of a point process approach to the top of the spectrum of $\Delta^{\rm d} + \xi$, with i.i.d. random potential ξ, in large boxes. We recommend [Res87] for an extensive background on extreme-value theory, and on point processes and their convergence. We refer to the eigenvalues $\lambda_k = \lambda_k(B)$ and orthonormal basis of eigenfunctions v_k of $\Delta^{\rm d}+\xi$ in a large box B with zero boundary condition as in (6.2).

What do the terms λ_k and $v_k(0)$ express? We indicated the answer already in Sect. 2.2.3, where we said that one crucial fact from Anderson localisation theory is the exponential decay of the eigenfunction v_k away from its localisation centre z_k. Hence, $v_k(0)$ should somehow measure the distance of z_k to the origin, i.e., it should be upper bounded against something like a negative exponential of $|z_k|$. Furthermore, the corresponding k-th largest eigenvalue λ_k should indeed be seen as the k-th largest *principal* eigenvalue in a small sub-box of B, and we will make this important identification throughout these heuristics. These sub-boxes are the local regions of very high values of the potential in the sense of an extreme-value statement, and they are far apart from each other and hence practically independent. Hence, the logarithm of the term $e^{t\lambda_k}|v_k(0)|$ is a difference between the 'gain' coming from a large local principal eigenvalue and the 'cost' for the local region being far away from the origin.

A control on all these values is achieved by means of a point process convergence of both the eigenvalues and the distances of the localisation centres to the origin towards a Poisson point process, after suitable rescaling and shifting. By this we mean a statement of the form

$$\sum_{k=1}^{|B_L|} \delta_{((\lambda_k(B_L)-a_L)b_L, z_k/L)} \stackrel{L\to\infty}{\Longrightarrow} \mathrm{PPP}(\mathbb{R}\times[-1,1]^d; \mu\otimes \mathrm{Leb}|_{[-1,1]^d}), \tag{6.11}$$

where $B_L = [-L,L]^d \cap \mathbb{Z}^d$, $\lambda_k(B_L)$ and v_k are the eigenvalues and eigenfunctions of $\Delta^{\rm d} + \xi$ in B_L, $z_k \in B_L$ are the localisation centres of v_k, and a_L and b_L are suitable norming constants, and μ is a suitable intensity measure on $\mathbb{R}$ for the limiting Poisson point process. We already anticipated that the limiting locations of the normalised localisation centres are homogeneous by putting the Lebesgue measure on $[-1,1]^d$ as the spatial part of the intensity measure.

From (6.11), we obtain the information that all the eigenvalues that are of order a_L, have gaps of order $1/b_L$ between successive ones of them and that each of them converges in distribution to some non-trivial explicit distribution; furthermore, their localisation centres have distances of order L from each other. Note that, a priori, all the other eigenvalues diverge in this limit either to $+\infty$ or to $-\infty$ and disappear in the limiting process. The choice of the scaling terms a_L and b_L determine in what part of the spectrum we derive non-trivial assertions.

In the application for the PAM, we are exclusively interested in the top of the spectrum, i.e., in $\lambda_1(B_L), \lambda_2(B_L), \dots$. In order to fish out these, we need to take a_L and b_L as suggested from the extreme-statistics behaviour of these random variables. A proper choice for a_L can be roughly guessed from the requirement

$$\mathrm{Prob}(\lambda_1(B_R) > a_L) \approx \frac{1}{|B_L|}, \qquad L \to \infty, \tag{6.12}$$

where B_R is a box (a 'microbox') in which the corresponding eigenfunction will have most of its mass. This requirement guarantees that we may expect approximately one microbox of radius R in B_L with local principal eigenvalue of size a_L with high probability, and a_L is maximal with this property. Having identified a_L by (6.12), one needs to find the second crucial scale b_L by the requirement that $\mathrm{Prob}(\lambda_1(B_R) \leq a_L + s/b_L)|B_L|$ converges to some non-trivial distribution function in $s \in \mathbb{R}$. Having found the right norming constants, (6.11) follows from standard techniques from the theory of extreme-value statistics. Also the identification of the limiting distributions of the rescaled eigenvalues, $(\lambda_k(B_L) - a_L)b_L \overset{L\to\infty}{\Longrightarrow} \gamma_k$, follows from standard devices; in general they have one of the three famous extreme-value distributions, the Weibull, Gumbel or Fréchet distribution.

Actually, the above explanation can serve only as a heuristic, as all the proofs that we know are not able to get sufficient control on $v_k(0)$, since this is not a quantity that can be expressed in terms of convergent objects in the light of the above point process convergence. Instead, one works with probabilistic tools, using the Feynman-Kac formula.

Strictly speaking, is not Anderson localisation that is exploited in the above heuristics, as this term applies by definition only in the $\mathbb{Z}^d$-setting, but not in large boxes, but there is no doubt that the localisation picture should persist also to this setting, and that for pretty many potential distributions.

In those cases in which the eigenfunctions are very strongly localised, i.e., on single sites with a delta-like shape, this localisation picture is not far from the standard case of just i.i.d. random variables instead of local eigenvalues. This is essentially the class (SP) and will be highlighted in Sect. 6.3.3. In the general case, it is not easy to separate the concentration regions oft the eigenfunctions from each other. One important such class, the class (DE), is presented in Sect. 6.3.2. Precise assertions for other classes do not seem to exist in the literature, see some remarks in Sect. 6.3.3.

6.3.2 *The Class (DE)*

In this section, we give details about the eigenvalue order statistics for the double-exponential distribution (the case (DE)), which is based on [BisKön16]. See Sect. 6.3.3 for the other potential classes.

Let the potential distribution lie in the class (DE), i.e., $\mathrm{Prob}(\xi(0) > r) \approx \exp\{-e^{r/\rho}\}$ for large $r \in \mathbb{R}$, see Remark 3.17, with some technical extra restrictions. Note that the maximum of the i.i.d. potential ξ in the box $B_L = [-L, L]^d \cap \mathbb{Z}^d$ is equal to $\rho \log\log |B_L| + o(1)$ as $L \to \infty$, as one can easily derive from the upper tails (see also [GärMol98]). Recall the characteristic variational formula χ and related information from Remark 3.17. Hence, $\lambda_1(B_L)$, the largest eigenvalue of $\Delta^{\mathrm{d}} + \xi$ in B_L, is $\rho \log\log |B_L| - \chi + o(1)$. Let us define a_L by the requirement that $\mathrm{Prob}(\lambda_1(B_{\log L}) \geq a_L) = 1/|B_L|$, see (6.12). (One has to pick the radius of the microbox divergent, since the full spectral description of the intermittent islands requires a fixed radius of arbitrarily large size to get arbitrary precision, hence we made the choice $R = \log L$.) Then $a_L = \rho \log\log |B_L| - \chi + o(1)$, and we can expect that, with high probability, there is one local region of radius $\log L$ in B_L whose local eigenvalue is roughly equal to a_L.

Now we can deduce much more information about all the top eigenvalues and eigenfunctions, $\lambda_k(B_L)$ and v_k, in B_L with zero boundary condition.

Theorem 6.4 (Eigenvalue order-statistics in case (DE)) *For each $L \geq 1$ there is a sequence $Z_1^{(L)}, Z_2^{(L)}, \dots$ of random sites in B_L such that, for any $R_L \to \infty$, in probability,*

$$\lim_{L\to\infty} \sum_{z : |z - Z_k^{(L)}| \leq R_L} |v_k(z)|^2 = 1, \qquad k \in \mathbb{N}. \tag{6.13}$$

Moreover, the law of the point process

$$\sum_{k \in \mathbb{N}} \delta_{\left((\lambda_k(B_L) - a_L)\log L,\, Z_k^{(L)}/L\right)} \tag{6.14}$$

converges weakly towards a Poisson point process on $\mathbb{R} \times [-1, 1]^d$ with intensity measure $\mathrm{e}^{-\lambda}\,\mathrm{d}\lambda \otimes \mathrm{Leb}(\mathrm{d}x)$.

Equation (6.13) states the concentration of the k-th eigenfunction in a R_L-box centred at some random location $Z_k^{(L)}$; this is the spectral concentration part of the assertion. Equation (6.14) is an example of (6.11) with $a_L = \rho \log\log |B_L| - \chi + o(1)$ and $b_L = \log L$. In particular, the eigenfunction localisation centres converge towards a standard Poisson process, after rescaling with L, and any two neighbouring eigenvalues have distance of order $1/\log L$ to each other. More precisely, the rescaled eigenvalue $(\lambda_1(B_L) - \rho \log\log L^d + \chi) \log L$ converges weakly towards a standard Gumbel variable, and all the other rescaled eigenvalues converge towards some explicit random variables that can be expressed in terms of i.i.d. copies thereof. Formulated in one compact statement, we have

$$\begin{aligned}&\big(\mathrm{e}^{-\frac{1}{\rho}(\lambda_1(B_L)-a_L)\log|B_L|},\dots,\mathrm{e}^{-\frac{1}{\rho}(\lambda_k(B_L)-a_L)\log|B_L|}\big)\\&\qquad\overset{L\to\infty}{\Longrightarrow}\ (E_1,E_1+E_2,\dots,E_1+\dots+E_k),\qquad k\in\mathbb{N},\end{aligned}\tag{6.15}$$

where $E_1, E_2, \dots$ are i.i.d. exponential with parameter one. Equivalently, the vector on the left tends in law to the first k points of a Poisson point process on $[0,\infty)$ with intensity one. In particular, the gaps between two subsequent eigenvalues $\lambda_k(B_L)$ and $\lambda_{k+1}(B_L)$ are in distribution equal to $1/\log L$ times some explicit random variables.

Let us give some heuristic explanation about some core steps of the proof of Theorem 6.4.

Step 1. The top eigenvalues are essentially local principal eigenvalues, and hence independent. Let us explain why the leading eigenvalues (i.e., those λ_k whose values are not far from the one of λ_1 in B_L) are extremely close to the principal eigenvalue in some small box. Indeed, pick some threshold $A > 0$ and put $U = \bigcup_{z\in B_L:\xi(z)\geq\lambda_1(B_L)-2A} B_R(z)$ for some radius $R > 0$, then U is the union of R-balls around the sites carrying highest potential values in B_L. Then, according to [BisKön16, Theorem 2.1], for any k such that $\lambda_k \geq \lambda_1 - A/2$, we have $|\lambda_k(B_L) - \lambda_k(U)| \leq 2d(1+\frac{A}{2d})^{1-2R}$, i.e., the difference between the k-th eigenvalue in B_L and the k-th eigenvalue in the (much smaller) set U is tiny for large R. (To see this, one shifts $\xi(z)$ for $z \in B_L \setminus U$ towards $-\infty$ and uses that the derivative of the eigenvalue with respect to $\xi(z)$, i.e.,

$$\frac{\partial}{\partial\xi(z)}\lambda_k(B_L) = \varphi_k(z)^2, \qquad z\in B_L. \tag{6.16}$$

is tiny.) Taking now into account that, with high probability, U decomposes into many components of bounded size [GärMol98, Corollary 2.10] and are far away from each other [GärKönMol07, Lemma 5.1], and knowing that the second-largest eigenvalue in such a component is bounded away from the principal one [BisKön16, Proposition 2.2], we see that each eigenvalue $\lambda_k(U)$ with sufficiently large k must be the principal one in one of these components. This argument also shows that the global top eigenvalues $\lambda_k(B_L)$ can be well approximated with local principal ones, and the latter are independent.

Step 2. The local principal eigenvalues satisfy an order statistics. For having this, one needs an assertion of the form

$$\mathrm{Prob}(\lambda_1(B_{\log L},\xi) \geq a_L + s/b_L) \sim \frac{\mathrm{e}^{-s}}{|B_L|} \text{ for any } s\in\mathbb{R}, \tag{6.17}$$

as $L \to \infty$, for some scale function b_L, where we recall that a_L was picked such that this holds for $s = 0$. If ξ is given precisely as the double-exponential distribution in (3.44), this implication is shown to be true with the value $b_L = \frac{d}{\rho}\log L$. The main point here is that the behaviour of this distribution under shifts $\xi \mapsto \xi + c$ is rather easily identified explicitly. This is the core of a proof not only of an eigenvalue order statistics in the domain of attraction of the Gumbel distribution, but even of the convergence of the point process of rescaled eigenvalues, together with the rescaled localisation centres of the corresponding eigenfunctions, towards an explicit Poisson point process. In particular, we know now that the gaps between any two subsequent eigenvalues behaves like $1/\log L$ times an explicitly known random variable.

Step 3. Eigenvalue gap implies eigenfunction decay. This a kind of argument that is well-known in Anderson localisation theory (see e.g. [Kir10]): the eigenfunction corresponding to an eigenvalue that has a notable, positive distance to all the other eigenvalues decays exponentially fast away from some site, its localisation centre. For proving this, in [BisKön16], an argument is employed that shows that the eigenfunction remains practically unchanged if the potential is shifted to $-\infty$ outside of a neighbourhood of the local island of sites that give extremely high potential values and carries some mass $\geq 1/2$ of the eigenfunction. For this, one first shows that the process

$$\Big(v_k(Y_n)\prod_{l=1}^{n}\frac{2d}{2d-\lambda_k(B_L)-\xi(Y_l)}\Big)_{n\in\mathbb{N}}$$

is a martingale, where $(Y_n)_{n\in\mathbb{N}}$ is a discrete-time simple random walk starting from the localisation centre. Since the quotient in the product lies in $(1,\infty)$ and is bounded away from 1 as long as Y runs outside the highest peaks, this property makes it quite easy to show that the eigenfunction v_k decays exponentially fast away from the area of high exceedances of ξ. Using this in (6.16), we see that the eigenvalue does not change if the potential is drastically changed there.

This ends our heuristic explanation of the proof of Theorem 6.4.

6.3.3 Other Classes of Potentials

In this section, we report on assertions about eigenvalue order statistics and point-process convergence for the eigenvalues and eigenfunction localisation centres for other classes of potential distributions.

Let us start with the most heavy-tailed ones. As we will see in Sect. 6.4.2, for several distributions in the class (SP) (i.e., for the heavier-tailed distributions of Example 1.14), the mass concentration phenomenon for the PAM in one island has been proved in the literature. For the heaviest-tailed distributions like Pareto, Weibull with parameter in $(0, 2)$ and the exponential distribution, no explicit formulation of an eigenvalue order statistics is necessary, as the intermittent islands are just singletons. The concentration in these singletons can be proved without help from spectral properties like shapes of eigenfunctions in their neighbourhoods, in contrast to the case (DE). However, an order statistics theorem like in (6.14) for the i.i.d. potential values instead of the eigenvalues is an important ingredient there, but was not in all that papers explicit. E.g., for the Pareto distribution, it turned out in [KönLacMörSid09] that the potential values, and hence the top eigenvalues, lie in the max-domain of a Fréchet distribution, but no explicit point process convergence was necessary, since only the three largest potential values needed to be looked at. The Pareto distribution is the heaviest distribution that can be used for the PAM (recall the condition in (1.5)), and it was the first one for which the concentration property of the PAM was proved.

However, the eigenvalue order statistics for heavy-tailed potential distributions, in the same vein as in Theorem 6.4 and without relation to the time-dependent problem (the PAM), have been derived in a series of papers [Ast08, Ast12, Ast13], together with detailed asymptotics in distribution for the distance of the normalised eigenfunctions from the delta-function in the localisation centre and much more detailed information. (See the recent extensive survey [Ast16].) They are also used as a basis for the proof of the mass concentration property for the PAM with Weibull distribution with parameter in $[2, \infty)$ in [FioMui14]. For this distribution (which is within the class (SP) closest to the class (DE)), the shape of the eigenfunctions in a certain neighbourhood of the concentration singletons must be taken into consideration, even though the neighbouring values of the eigenfunctions vanish quickly.

For the two interesting cases (AB) and (B), to the best of our knowledge, neither an assertion on the eigenvalue order statistics, nor a concentration property of the PAM has been proved yet, and we consider this as an interesting research project for the future, at least for some prominent representatives. We conjecture that in both these cases the principal eigenvalues lie in the max-domain of a Gumbel distribution. However, this should hold in the case (B) only for $\gamma > 0$ (recall the notation in Remark 3.19), while the case $\gamma = 0$ should lead to the Weibull distribution. One important difficulty that one has to overcome in the case (B) is that the change from $\{\lambda_k(B_{\log L}) \geq a_L\}$ to $\{\lambda_k(B_{\log L}) \geq a_L + s/b_L\}$ does, in contrast to the double-exponential distribution, not predominantly come from making each single potential site greater by the amount s/b_L, but from making the radius of the ball larger in which the potential gives the main contribution. For (AB), a combination of these two effects will be in order. However, we believe that several tools from [BisKön16] can be used or easily adapted.

Let us also remark that, to the best of our knowledge, there is no detailed assertion in the literature about point process convergence for the eigenvalues for any Gaussian random field in large boxes of $\mathbb{R}^d$, not even under strong regularity assumptions.

6.3.4 Relation to Anderson Localisation

Let us make some remarks on similar eigenvalue expansions for the Anderson Hamiltonian that were achieved in the community of random Schrödinger operators. Indeed, the joint distribution of the eigenvalues and the concentration centres of the eigenfunctions is of great interest, as it gives important information about Anderson localisation, see Sect. 2.2.3. In this community, the interest in such information is not restricted to the top of the spectrum, but extends to all parts of the spectrum where Anderson localisation is known to hold. Consequently, statements like the one in (6.11) do typically have nothing to do with extreme-value analysis, and the norming constants a_L and b_L are picked according to other requirements than the one in (6.12), as one mostly works inside the spectrum, i.e., away from the boundaries. Consequently, the order $1/b_L$ of the gap size is then $1/|B_L|$, in contrast to the much larger gaps at the spectral boundary (e.g., $1/\log|B_L|$ for the double-exponential distribution). Furthermore, the limiting Poisson process has a different intensity measure, since it does not come from extreme values.

An early example is [Mol81], where the eigenvalues of the one-dimensional Anderson Hamiltonian on $\mathbb{Z}$ are shown to have a Poisson process structure. An important progress is made in [Min96], where this kind of assertion is extended to the d-dimensional setting in $\mathbb{R}^d$. The main value of [Min96] is the introduction of a flexible estimate that establishes the existence of a gap between two subsequent eigenvalues, an estimate that is now called the *Minami estimate*. The first result on the convergence of point processes of both the eigenvalues and the concentration centres of the eigenfunctions is [KilNak07]; see also [Nak07]. The currently strongest available results are in [GerKlo14] and [GerKlo13], where [GerKlo14] works in the bulk of the spectrum and [GerKlo13] close to the top; see also [GerKlo11].

The latter two works consider much more general random operators $\mathcal{H}$ than just the Anderson operator $\Delta^{\rm d} + \xi$ on $\mathbb{Z}^d$ or $\Delta + V$ on $\mathbb{R}^d$, and they assume that the potential distribution has a bounded density and that Anderson localisation holds in the spectral interval considered. Furthermore, they make a couple of assumptions on the validity of Wegner and Minami estimates, which are known to hold for large classes of random operators $\mathcal{H}$. They pick a growing number of eigenvalues in the interval considered and corresponding eigenfunction centres and show that their point process, after 'unfolding', converges towards a standard Poisson process with intensity measure $\mathrm{d}\lambda \otimes \mathrm{d}x$, i.e., the Lebesgue measure both in spectrum and space.

More precisely, they do not look at a rescaling $(\lambda_k(B_L)-a_L)b_L$ for box-depending quantities a_L and b_L, but on the *unfolded eigenvalues* $[\mu(\lambda_k(B_L)) - \mu(\lambda_0)]|B_L|$, where $\mu\colon\mathbb{R} \to [0,1]$ is the integrated density of states (IDS), see Sect. 2.2.6, and

λ_0 is a certain value in the spectrum of the global operator $\mathcal{H}$ that satisfies some additional properties. Since λ_0 is assumed to lie in the interior of the support of the IDS, also [GerKlo13] makes assertions only for eigenvalues that are substantially away from the boundary of the spectrum (however, it contains also a restricted assertion precisely at the boundary for the one-dimensional operator $\mathcal{H} = \Delta^{\mathrm{d}} + \xi$). All the assertions proved in [GerKlo14] and [GerKlo13] do not come from any kind of maximisation and therefore have nothing to do with extreme-value analysis, nor there are assertions about the shape of the potential inside the islands. A comparison to the approach described above is not immediately clear.

6.4 Concentration in One Island

Let us now come to the strongest assertion on intermittency that one can think of: the contribution to the total mass $U(t)$ coming from the solution u of the PAM in (1.1) and (1.2) in the complement of *just one island* is negligible with respect to the one coming from this island itself. This is the assertion in (6.1) for $n_t = 1$. Such a strong assertion has been proved for several distributions in the class (SP) (i.e., with concentration in one single site) and for the class (DE) containing the double-exponential distribution. However, such a statement is lacking yet for both the cases (AB) and (B) and also for the spatially continuous setting. (However, note that it has been proved for the PAM in a special random environment with Weibull distributed potential, see Sect. 7.9.2.)

It was a certain change of paradigms around 2006 to look at the PAM with random potential in a class that has no finite exponential moments, as the solution u to the PAM then does not have any moment. Therefore, moment intermittency defined by (1.10) cannot be considered and these potentials a priori do not fall into any of the four potential classes introduced in Sect. 3.4 (but belong in spirit to the class (SP)). However, starting with [HofMörSid08], it became quickly clear that such potentials are highly interesting and push the investigation of the PAM much further, as they offer the possibility to study the geometric picture of intermittency in great detail, while posing tractable, but still considerable, mathematical difficulties. They are easier to handle, as the islands are singletons, and the spectral landscape can be treated in some respects like an i.i.d. field. Later, additional efforts were invested to manage also the class (DE), where the islands carry interesting structure.

We are going to outline the case of a double-exponentially distributed potential in Sect. 6.4.1 and the Pareto distribution and other distributions in the class (SP) in Sect. 6.4.2. We are working here with convergence in distribution, but see Remark 6.7 for almost-sure versions. We remind on [Res87] as an excellent source of general information on extreme-value theory and point process convergence.

6.4.1 The Class (DE)

We consider an i.i.d. potential ξ in the class (DE) (see Remark 3.17) with parameter $\rho \in (0,\infty)$, i.e, with upper tails like $\mathrm{Prob}(\xi(0) > r) \approx \exp\{-\mathrm{e}^{r/\rho}\}$ for any large $r \in \mathbb{R}$, and with some mild technical restrictions. We follow [BisKönSan16]. Recall the notation from Theorem 6.4, in particular the Poisson point process on $\mathbb{R} \times \mathbb{R}^d$ with intensity measure $\mathrm{e}^{-\lambda}\,\mathrm{d}\lambda \otimes \mathrm{Leb}(\mathrm{d}x)$, which we now want to write as $(\gamma_i, x_i)_{i\in\mathbb{N}}$; note that we extend it here to the entire space $\mathbb{R} \times \mathbb{R}^d$. For $\theta \in (0,\infty)$, we introduce the function

$$\psi_\theta(\gamma, x) = \gamma - \frac{|x|}{\theta}, \qquad \gamma \in \mathbb{R}, x \in \mathbb{R}^d,$$

and denote by $(\gamma_\theta^*, x_\theta^*)$ the pair that maximises $\psi_\theta(\gamma_i, x_i)$ over all the $i \in \mathbb{N}$. (It is standard to see that, for almost all θ, this is well-defined and unique.) Then we can formulate the asymptotic concentration property of u in just one single island as follows.

Theorem 6.5 (One-island concentration in the case (DE)) *There is a $\mathbb{Z}^d$-valued process $(Z_t)_{t\in(0,\infty)}$ such that the following holds.*

(i) For any $\delta \in (0,1)$, there is $R \in \mathbb{N}$ satisfying

$$\liminf_{t\to\infty} \frac{1}{U(t)} \sum_{z:|z-Z_t|\le R} u(t,z) \ge 1-\delta \qquad \text{in probability.}$$

(ii) with $d_t = \rho/d \log t$ and $L_t = dt/\rho \log t \log\log\log t$,

$$\left(\frac{\frac{1}{\theta t}\log U(\theta t) - a_{L_t}}{d_t}, \frac{Z_{\theta t}}{L_t}\right)_{\theta\in(0,\infty)} \overset{t\to\infty}{\Longrightarrow} \left(\psi_\theta(\gamma_\theta^*, x_\theta^*), x_\theta^*\right)_{\theta\in(0,\infty)}, \tag{6.18}$$

where $\Longrightarrow$ denotes weak convergence on every compact time interval $\subset (0,\infty)$ in the Skorohod space.

(iii) In particular, for each $\theta > 0$, $(\frac{1}{\theta t}\log U(\theta t) - a_{L_t})/d_t$ converges in law to a Gumbel random variable with scale 1 and location $d\log(2\theta)$, while $Z_{\theta t}/L_t$ converges in law to a random vector in $\mathbb{R}^d$ with Lebesgue density $x \mapsto (2\theta)^{-1}\mathrm{e}^{-|x|/\theta}$.

Hence, the total mass essentially comes from a single island, and the larger the island is taken, the more percentage of the total mass is captured. This is (6.1) with $n_t = 1$. Furthermore, the island is at distance $\asymp t/(\log t \log\log\log t)$ from the origin, and the Poisson point process introduced in Theorem 6.4 describes both the location of the island (i.e., of its centre point Z_t) and the principal eigenvalue of $\Delta^{\rm d} + \xi$ in the island, via the approximations $\frac{1}{t}\log U(t) \sim \lambda_1(B_R(Z_t),\xi) - |Z_t|/r_t$ and $a_{r_t} \approx \max_{B_{r_t}} \xi - \lambda_1(B_R(Z_t),\xi) \approx \rho\log\log t - \chi$ for $t \to \infty$, followed by $R \to \infty$; see Remark 6.6 for more details. Assertion (iii) is standard in the theory of Poisson point processes; it shows that the eigenvalue-order statistics lies in the Gumbel class.

Remark 6.6 (Relation with eigenvalue order statistics) The spatial scale $L_t \asymp t/\log t\log\log\log t$ comes from an optimisation of the travel distance of the trajectory and the local eigenvalue in the Feynman-Kac formula, using the knowledge that we have from Theorem 6.4. Indeed, inserting the strategy that the path needs s time units to reach an optimal island B at distance L from the origin and with eigenvalue λ and then remains $t-s$ time units in that island, gives the approximation

$$\mathbb{E}_0\Big[{\rm e}^{\int_0^t \xi(X_r)\,{\rm d}r}\mathbb{1}\{|X_s| \asymp L\}\mathbb{1}\{X_r \in B\,\forall r \in [s,t]\}\Big] \approx \mathbb{P}_0(|X_s| \asymp L){\rm e}^{(t-s)\lambda} \approx {\rm e}^{-L\log(L/s)}{\rm e}^{(t-s)\lambda}.$$

(For the path probability approximation we used (3.5), as the optimal L should be picked much larger than s.) Now we optimise over $s \in [0,t]$, noting that λ is one of the local principal eigenvalues λ_k, which behaves like $a_L + \gamma_k/\log L \approx \rho\log\log L - \chi + \gamma_k/\log L$, with one point γ_k of the limiting Poisson process. Hence, we choose $s = L/\lambda_k \approx L/a_L$, assuming (and later justifying) that this is $\ll t$. After applying some elementary approximations, we see that the function

$$\Psi_{L,t}(\lambda,z) = \frac{t}{r_L}(\lambda - a_L)\log L - \frac{|z|}{L}, \qquad \text{where } r_L = L\log L\log\log\log L, \tag{6.19}$$

describes the exponential rate on the scale r_L from the contribution to the Feynman-Kac formula coming from such a path behaviour. Now we pick L_t according to $r_{L_t} \asymp t$, which gives a balance between the eigenvalue rescaling and the spatial terms, makes both terms running on a finite scale, and implies that $L_t \asymp t/\log t\log\log\log t$. Recall the point process of eigenvalues and localisation centres of the eigenfunctions from Theorem 6.4, then, formally combining the limit $L \to \infty$ with the limit $t \to \infty$, we see that one should have

$$\Big(\Psi_{L_t,t}\big(\lambda_k(B_{L_t},\xi), Z_k^{(L_t)}\big)\Big)_{k\in\mathbb{N}} \overset{t\to\infty}{\Longrightarrow} \big(\psi_1(\gamma_k,x_k)\big)_{k\in\mathbb{N}}.$$

Picking the maximum over k yields Theorem 6.5(ii). This heuristics explains only the concentration property within the centred box of radius of order $t/\log t\log\log\log t$. It is relatively easy to show that the outside of the box with radius $t\log^2 t$ is negligible, so some technical work is to be done for showing that also the sphere between these two is negligible. ◇

6.4.2 The Class (SP)

Now let us turn to the class (SP) of random potentials, see Remark 3.16. We begin with the heaviest tails, the case of a Pareto-distributed potential, and follow [HofMörSid08, KönLacMörSid09, MörOrtSid11]; see also the survey [Mör11] on this special aspect of the PAM.

The study of the PAM with thick-tailed potentials was initiated in [HofMörSid08], where almost sure and distributional limit theorems for the total mass $U(t)$ are derived for the Weibull and the Pareto case. We discuss here the Pareto distribution $\mathrm{Prob}(\xi(0) > r) = r^{-\alpha}$ for $r \in [1, \infty)$ with some $\alpha \in (d, \infty)$ (recall that the parameter α must exceed d to satisfy (1.5)). Here, it is proved that

$$\Big(\frac{t}{\log t}\Big)^{\frac{-d}{\alpha-d}} \times \frac{1}{t}\log U(t) \overset{t\to\infty}{\Longrightarrow} Y, \qquad \text{where } \mathbb{P}(Y \leq y) = \exp\{-\theta y^{d-\alpha}\}, \tag{6.20}$$

and θ is some explicit constant. Furthermore, explicit almost sure liminf and limsup results for the logarithm of $\frac{1}{t}\log U(t)$ are derived. Note that the limiting distribution in (6.20) is the Fréchet distribution, another one of the three famous limiting distributions for the maximum of i.i.d. random variables. Hence, the assertion of (6.20) is very much in line with the understanding that all the leading eigenfunctions in the expansion (2.18) are delta-like functions, and $\frac{1}{t}\log U(t)$ is approximately equal to the maximum of a large number of i.i.d. Pareto-distributed random variables.

In [KönLacMörSid09], techniques from [GärKönMol07] (in particular, the device outlined in Remark 6.2) were added to prove the concentration property. More precisely, it was proved that there is a stochastic process $(Z_t)_{t\in(0,\infty)}$ in $\mathbb{Z}^d$ such that

$$U(t) \sim u(t, Z_t) \qquad \text{as } t \to \infty \text{ in probability,} \tag{6.21}$$

which is (6.1) with $n_t = 1$. The fact that the difference between the largest and the second-largest of the potential values in the box is huge, is helpful in the proof at some places. An informal description of the site Z_t is as follows. Consider the function

$$\Psi_t(z) = \xi(z) - \frac{|z|}{t}\log\frac{|z|}{2det}, \qquad z \in \mathbb{Z}^d, t > 0, \tag{6.22}$$

then $e^{t\Psi_t(z)}$ is roughly equal to the contribution to the Feynman-Kac formula in (2.2) coming from a path that quickly runs to the site z and stays in z for the rest of the time until t (analogously to Remark 6.6). Indeed, the first term is the potential value that is attained for $\approx t$ time units, and the second is the probability to go for a distance $|z|$ in $\approx o(t)$ time units. (The function Ψ_t plays a similar rôle here as the function $\Psi_{L,t}$ in (6.19) for the double-exponential distribution.) Then Z_t is defined as

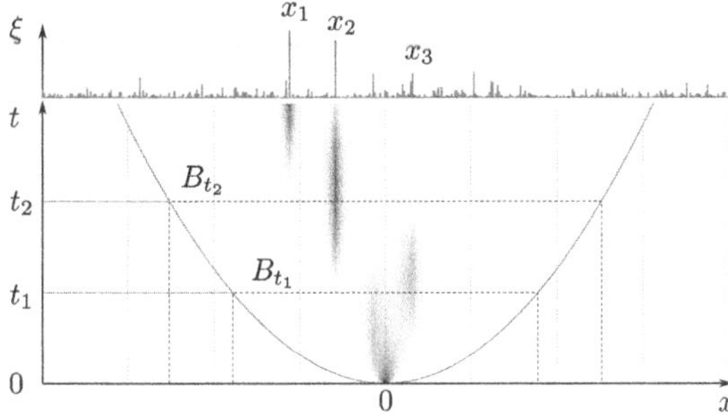

Fig. 6.1 One realisation of a one-dimensional Pareto-distributed potential and some information on the time-evolution. At the respective times, the parabola shows the horizon that the mass flow sees, and the *shaded areas* indicate the dominant sites. By time just before t_1, the peak just left of the origin is dominant, afterwards the peak in x_3 takes over until time just before t_2, then the peak in x_2 takes over, then the one in x_1 until time t. Observe that the dominance is transferred continuously from one peak to the next one during some non-trivial (random) time stretches. During some part of them, there are two dominant sites

the site that maximises Ψ_t. In particular, $\Psi_t(Z_t) = \max_{z\in\mathbb{Z}^d} \Psi_t(z) \approx \frac{1}{t}\log U(t)$. We discuss the entire process $(Z_t)_{t\in[0,\infty)}$ in Sect. 6.5.

Remark 6.7 (Almost sure concentration: a two-cities theorem) The asymptotics in (6.21) cannot be true almost surely. In this case, t would be a random time and would also sooner or later attain a value that lies in a time interval during which the dominant potential peak wanders from one location to another one. Such phases of wandering of the overwhelming mass from one 'city' to the next one occur, since the horizon increases as t increases, and the maximisation of the field takes place over larger and larger areas, see Fig. 6.1. However, in [KönLacMörSid09] it is proved that the main mass is concentrated in no more than *two* sites at any large time t, almost surely. This interpretation inspired the title of the paper [KönLacMörSid09]. ◇

Remark 6.8 (Exponential distribution) Another interesting potential distribution that turns out to phenomenologically lie in the class (SP), is the *exponential distribution*, $\mathrm{Prob}(\xi(0) > r) = \mathrm{e}^{-r}$ for $r \in (0,\infty)$. This distribution is considered in [LacMör12], and it is found that a concentration property in one single site takes place as well. First, like for the Pareto distribution in [HofMörSid08], some distributional and almost sure liminf and limsup results for $\frac{1}{t}\log U(t)$ are given in [LacMör12]. Furthermore, it is shown that the point process

$$\frac{1}{U(t)}\sum_{z\in\mathbb{Z}^d} u(t,z)\delta_{z/L_t} \qquad \text{with } L_t = \frac{t}{\log\log t},$$

converges towards δ_Y, where Y is an $\mathbb{R}^d$-valued random variable with i.i.d. coordinates with exponential distribution with uniform random sign. ◇

Remark 6.9 (Weibull distribution) For the Weibull distribution with tails $\mathrm{Prob}(\xi(0) > r) = \mathrm{e}^{-r^\gamma}$, the assertion in (6.21) with a random process Z_t of order $t(\log t)^{1/\gamma-1}/\log\log t$ was proved in [SidTwa14] for $\gamma \in (0,2)$, and in [FioMui14] for $\gamma \in [2,\infty)$. The latter case is technically more involved, as one has to take into account the local principal eigenvalue in a certain *non-trivial* box (whose radius depends on γ only) around a local potential maximiser. Actually, the neighbouring sites also have very large values, which are quantified in [FioMui14] as $\xi(Z_t + z) = (d\log t)^{(1-2|z|/(\gamma-1))_+/\gamma}$. ◇

6.5 Ageing and Time-Correlations

The ultimate goal in the study of the PAM is of course the description of the heat flow through the random potential as a stochastic process in time, i.e., the description of the entire process $(u(t,\cdot))_{t\in(0,\infty)}$ of solutions for one realisation of the potential ξ. So far we reported exclusively on results about snapshots of the model at late times, but now we go into properties of the PAM that depend on the time-evolution, i.e., on the observation of the process at several times. In general, already formulations of such properties are rather cumbersome (not to mention proofs). However, as we reported on in Sect. 6.4, we are nowadays in the comfortable situation that we can convincingly characterise the main characteristics of the solution in terms of just the location Z_t of the single island in which the solution $u(t,\cdot)$ is concentrated. This gives us the key for formulating ageing properties of the PAM both in terms of the process $(Z_t)_{t\in(0,\ infty)}$ and in terms of the normalised solution $u(t,\cdot)/U(t)$.

Generally speaking, *ageing* is the phenomenon that the most prominent, drastic changes of the system occur after longer and longer time periods or after shorter and shorter time periods. Hence, an observer is able to say how old the system is, if he/she can measure the time period that elapses between two such changes, or he/she can estimate the time until the next such change takes place, if he/she knows at what time the system started to evolve. For the PAM, the most relevant drastic changes are the jumps of the concentration centre, i.e., the jumps of the location of the local region with the best compromise between the size of the principal eigenvalue and its distance to the origin, or the jumps of the location of the dominant peak in the landscape determined by the solution $u(t,\cdot)$. See Fig. 6.1 for an illustration of this phenomenon; Fig. 6.2 shows the relation with the leading eigenfunctions, as seen in the eigenfunction expansion in (6.2).

Detailed descriptions of ageing in various senses have been given for most of the potential distributions for which one has derived the one-island concentration property, that is, for the potential classes (DE) and (SP); see Sect. 6.4. We give an account on them below. Let u be the solution to the PAM in (1.1), (1.2) and $U(t)$ its total mass at time t.

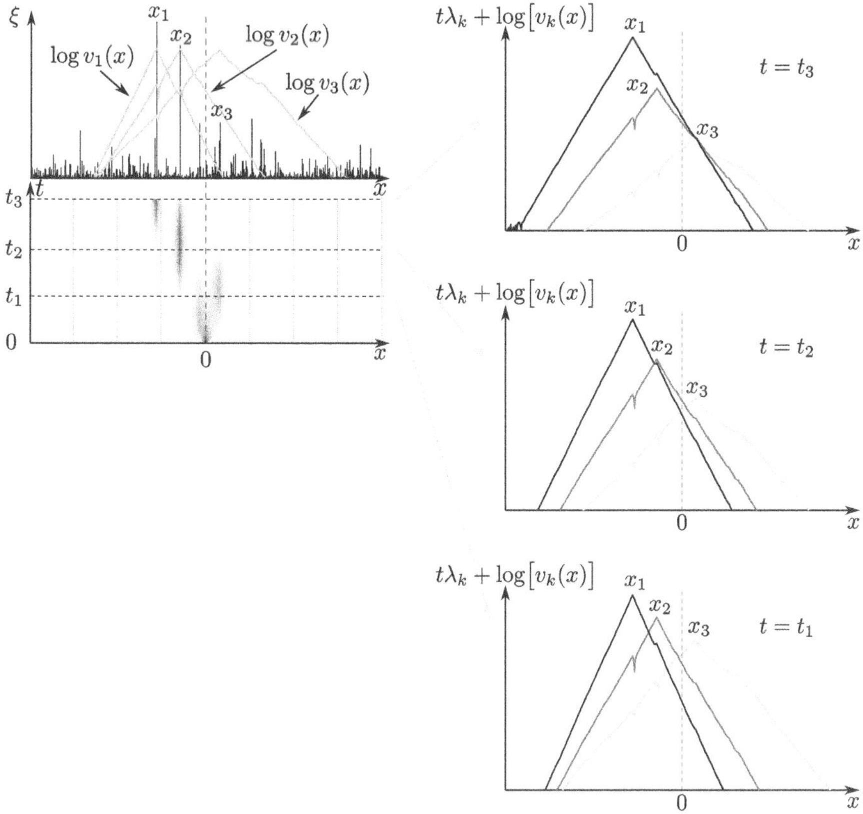

Fig. 6.2 Same realisation of a ξ as in Fig. 6.1. The logarithms of the three leading eigenfunctions are added, which decay linearly away from the localisation centres x_1, x_2 and x_3. In the three figures on the right, the three functions $x \mapsto t\lambda_k + \log v_k(x)$, $k = 1, 2, 3$, are depicted at times t_1, t_2 and t_3. Their values at $x = 0$ are decisive for the question which one dominates, as one sees in the eigenvalue expansion in (6.2)

6.5.1 The Class (DE)

As in Sects. 6.3.2 and 6.4.1, let ξ be an i.i.d. random potential in the class (DE). Recall all the notation from there, in particular the process $(Z_t)_{t\in(0,\infty)}$ of concentration loci of the solution u. For the limiting process $(x^*_\theta)_{\theta\in(0,\infty)}$, we denote by

$$\Theta := \inf\{\theta > 0\colon x^*_{1+\theta} \neq x^*_1\} \tag{6.23}$$

the time lag that elapses after time one until the next jump; Θ is positive and finite almost surely. We are citing from [BisKönSan16].

Theorem 6.10 (Ageing for the concentration locus and for the solution)

(i) For any $s > 0$,

$$\lim_{t\to\infty}\text{Prob}\big(Z_{t+\theta t} = Z_t\ \forall\theta\in[0,s]\big) = \lim_{t\to\infty}\text{Prob}\big(Z_{t+st} = Z_t\big)$$
$$= \text{Prob}\big(x^*_{1+\theta} = x^*_1\ \forall\theta\in[0,s]\big) = \text{Prob}\,(\Theta > s)\,. \qquad (6.24)$$

(ii) For any $\varepsilon\in(0,1)$,

$$\frac{1}{t}\inf\Big\{s>0\colon \sum_{z\in\mathbb{Z}^d}\Big|\frac{u(t+s,z)}{U(t+s)} - \frac{u(t,z)}{U(t)}\Big| > \varepsilon\Big\} \overset{t\to\infty}{\Longrightarrow} \Theta. \qquad (6.25)$$

That is, the trajectory of the concentration locus $t\mapsto Z_t$ makes jumps after time lags of order of the time that has already elapsed at the time of the observation: the later the observation is, the longer these time lags are. The normalised solution $t\mapsto u(t,\cdot)/U(t)$ makes notable jumps also after these time lags; the proof shows that their ℓ^1-distance is close to 0 for long time (more precisely, for Θt time units, when the observation starts at time t) and then makes a jump of size ≈ 1 within $o(t)$ time units.

6.5.2 *The Class (SP)*

A detailed description of ageing properties for the case of the Pareto distribution has been carried out in [MörOrtSid11], i.e., in the most heavy-tailed distribution of the class (SP). This was the first result of that kind for the PAM in the literature; see also the survey [Mör11]. In [MörOrtSid11], the description of the entire process $(Z_t)_{t\in(0,\infty)}$ of localisation sites Z_t at time t is identified as follows. It is proved that there is a (time-inhomogeneous) Markov process $(Y_t^{(1)}, Y_t^{(2)})_{t\in(0,\infty)}$ in $\mathbb{R}^d\times\mathbb{R}$ such that

$$\Big(\Big(\frac{\log t}{t}\Big)^{\frac{\alpha}{\alpha-d}} Z_{\theta t}, \Big(\frac{\log t}{t}\Big)^{\frac{d}{\alpha-d}}\xi(Z_{\theta t})\Big)_{\theta\in(0,\infty)} \overset{t\to\infty}{\Longrightarrow} \Big(Y_t^{(1)}, Y_\theta^{(2)} + \frac{d}{\alpha-d}|Y_\theta^{(1)}|\Big)_{\theta\in(0,\infty)}.$$

Here $Y_\theta^{(1)}$ and $Y_\theta^{(2)}$, after rescaling, are the maximizer and next-to maximiser of Ψ_t defined in (6.22). We also see that $(\frac{\log t}{t})^{\frac{\alpha}{\alpha-d}} Z_t$ converges in distribution to $Y = Y_1^{(1)}$, hence, the distance to the origin has the order $\gg (\frac{t}{\log t})^{\frac{\alpha}{\alpha-d}}$, which is much larger than in the case of potentials with finite exponential moments.

Even though not made explicit in [KönLacMörSid09], the point process consisting of the rescaled potential values and their locations in a large box converges towards an explicit Poisson point process, and [MörOrtSid11] contains also an explicit description of $(Y_\theta^{(1)}, Y_\theta^{(2)})_{\theta\in(0,\infty)}$ in terms of this process and illustrates this in terms of figures. This is analogous to Theorem 6.5; see also [Mör11]. For an animation of this process, see the homepage http://people.bath.ac.uk/maspm/animation_ageing.pdf.

Furthermore, it is shown there that $t \mapsto Z_t$ ages in the same way as in the case (DE), which we formulated in Theorem 6.10. More precisely, [MörOrtSid11, Theorem 1.1] says that, for any $\theta \in (0,\infty)$ and every sufficiently small $\varepsilon > 0$,

$$\begin{aligned}\lim_{t\to\infty} \mathrm{Prob}\Big(\sup_{s\in[t,t+t\theta]}\sup_{z\in\mathbb{Z}^d}\Big|\frac{u(t,z)}{U(t)}-\frac{u(s,z)}{U(s)}\Big|<\varepsilon\Big)\\ =\lim_{t\to\infty}\mathrm{Prob}\Big(\sup_{z\in\mathbb{Z}^d}\Big|\frac{u(t,z)}{U(t)}-\frac{u(t+t\theta,z)}{U(t+t\theta)}\Big|<\varepsilon\Big)=I(\theta),\end{aligned} \tag{6.26}$$

where $I(\theta)$ is a number in $(0,1]$ with boundary behaviours given by $\theta(1-I(\theta)) \sim C_1$ for $\theta \downarrow 0$ and $I(\theta) \sim \theta^{-d} C_2$ as $\theta \uparrow \infty$ for some $C_1, C_2 \in (0,\infty)$. Recall from (6.21) that $u(t,\cdot)/U(t)$ is essentially a delta-function at Z_t, and the rescaling of Z_t is a jump process.

Also for other potential distributions in the class (SP), versions of Theorem 6.10 have been proved; see Theorem 1.3 in [SidTwa14] for the Weibull distribution with parameter $\in (0,2)$ and Theorem 1.3(c) and Corollary 1.4 in [FioMui14] for the Weibull distribution with parameter $\in [2,\infty)$; see Remark 6.9.

Remark 6.11 (Almost-sure asymptotics for the ageing time lags) Consider the waiting time $R(t) = \sup\{s \in [0,\infty)\colon Z_t = Z_{t+s}\}$, at time t, until the next jump of the process Z, then (6.26) says that the limiting distribution of $R(t)/t$ has distribution function $1 - I$. In addition, [MörOrtSid11, Theorem 1.3] says that, almost surely, for any non-decreasing function $h\colon (0,\infty) \to (0,\infty)$,

$$\lim_{t\to\infty}\frac{R(t)}{th(t)}=\begin{cases}0 & \text{if } \int_1^\infty \frac{\mathrm{d}t}{th(t)^d}<\infty,\\ \infty & \text{if } \int_1^\infty \frac{\mathrm{d}t}{th(t)^d}=\infty.\end{cases} \tag{6.27}$$

◇

Remark 6.12 (Correlation ageing) One of the most popular definitions of ageing is in terms of correlations. A process $Y = (Y_t)_{t\in[0,\infty)}$ is said to satisfy *correlation ageing* if, for some scale functions $s_1(t)$ and $s_2(t)$, tending to infinity as $t \to \infty$,

$$\lim_{t\to\infty}\frac{\mathrm{cov}\,(Y(t),Y(t+s_i(t)))}{\sqrt{\mathrm{Var}(Y(t))\mathrm{Var}(Y(t+s_i(t)))}}=\begin{cases}0 & \text{if } i=1,\\ 1 & \text{if } i=2.\end{cases}$$

It appears unclear in which way this definition makes any intuitive sense for Y equal to the process of total masses, $(U(t))_{t\in(0,\infty)}$, e.g., even though a proof of correlation ageing seems to be within reach for a number of interesting potentials. However, it does make a lot of sense to study time correlations of the solution to the PAM for many choices of two time instants, and this line of research has been initiated in [GärSch11]. ◇

Chapter 7
Refined Questions

Let us survey a number of questions around the PAM that go beyond the basic questions that we have treated so far. In Sect. 7.1 we show what refined techniques can say about deeper analysis of the moment asymptotics of the total mass. Correlated potentials are considered in Sect. 7.2. In Sect. 7.3, we multiply the potential with a small t-dependent prefactor and examine how the concentrated behaviour is turned into some homogenised one. In Sects. 7.4 and 7.5 we discuss connections between the research on the PAM and on the upper tails of the random walk in random scenery and of general self-attractive functionals of the local times, respectively. In Sect. 7.6 we make a few remarks on general self-attractive path measures, in Sect. 7.7 we show how to interpolate between the moment asymptotics and the almost-sure asymptotics of the total mass, in Sect. 7.8 we report on research on the PAM with other random paths than the simple random walk. Results for the PAM in random environment (i.e., when the simple random walk is replaced by some random walk in random environment) are described in Sect. 7.9. In Sect. 7.10 we briefly characterize the relationship of the PAM with another model of high interest, the directed polymers in random environment, and in Sect. 7.11 we discuss some recent research on branching random walks in random environment that was inspired by the research on the PAM.

7.1 Beyond Logarithmic Asymptotics, and Confinement Properties

The asymptotics of the moments of the total mass of the solution of the PAM in Theorem 3.13 describes just the two leading terms. Here we discuss what can be said about the next terms. This is intimately connected with a closer analysis of the behaviours of those realisations of the path and of the potential that give the main contribution to the moments, i.e., with extensions of the confinement properties

W. König, *The Parabolic Anderson Model*, Pathways in Mathematics,
DOI 10.1007/978-3-319-33596-4_7

that we briefly mentioned in Remarks 3.8 and 3.14. There, we pointed out that the relevant path respectively potential is (after some rescaling) attracted to the set of all shifts of the minimiser of the characteristic variational formula. Here, we even characterise the shift to which it is attracted.

To approach the problem and the way of thinking, we first describe it in the simplest situation: for the real line instead of function spaces.

Remark 7.1 (The Laplace method) First we recall, on a heuristic level, the well-known (refined) Laplace method in a simple example. The large-t asymptotics of the integral $\int_0^1 \mathrm{e}^{tf(x)}\,\mathrm{d}x$ for some continuous function $f\colon [0,1] \to \mathbb{R}$ can be described just as $\mathrm{e}^{t(\max_{[0,1]} f + o(1))}$. This is sometimes called the (weak version of the) *Laplace principle*. But one can also give much more precise asymptotics of the integral, and one can closer describe the set of x that give the biggest contribution to the integral, under more restrictive assumptions on f concerning its shape close to the maximiser(s). Let us assume that f possesses precisely one minimiser $x^* \in (0,1)$ and that f is twice continuously differentiable in a neighbourhood of x^*. Then a Taylor expansion shows, for $x \to x^*$, that

$$\begin{aligned} f(x) &= f(x^*) + (x-x^*)f'(x^*) + \frac{1}{2}(x-x^*)^2 f''(x^*)(1+o(1)) \\ &= \max_{[0,1]} f - \frac{(x-x^*)^2}{2\sigma^2}(1+o(1)), \end{aligned}$$

where $\sigma^2 = -1/f''(x^*) \in (0,\infty)$. Using this for x in a $O(1/\sqrt{t})$-neighbourhood of x^* gives for the integral the more precise asymptotics

$$\begin{aligned} \int_0^1 \mathrm{e}^{tf(x)}\,\mathrm{d}x &\approx \mathrm{e}^{t\max_{[0,1]} f} \int_{x^*-R/\sqrt{t}}^{x^*+R/\sqrt{t}} \mathrm{e}^{-t\frac{(x-x^*)^2}{2\sigma^2}}\,\mathrm{d}x \\ &= \mathrm{e}^{t\max_{[0,1]} f} \int_{-R}^{R} \mathrm{e}^{-\frac{y^2}{2\sigma^2}}\,\frac{\mathrm{d}y}{\sqrt{t}} \\ &\approx \mathrm{e}^{t\max_{[0,1]} f} \sqrt{\frac{2\pi\sigma^2}{t}}, \end{aligned} \tag{7.1}$$

where R is a large auxiliary parameter that is sent to ∞ at the end of the proof. Indeed, the asymptotics in (7.1) are precise up to equivalence, i.e., up to a factor of $1+o(1)$ in the limit $t \to \infty$. The abstract idea of the above is that, after extracting the leading term $tf(x^*)$ in the exponent, the density $tf(x) - tf(x^*)$ is absorbed in the change of the variable x into $t(x-x^*)^2$, which is of finite order for those x that really contribute.

The asymptotics in (7.1) are known as the (strong version of the) *Laplace principle*. Besides that this method brings the second term of the asymptotics to the surface, it also specifies the region that gives the main contribution to the integral of e^{tf}, namely an interval of radius $R/\sqrt{t}$ around the maximiser x^*, and it gets the more precise the larger R is picked. The decisive inputs in the method are the uniqueness of the minimizer of the characteristic variational problem (here x^* for $\max f$) and a smooth behaviour and non-trivial curvature of the functional (here f) in a neighbourhood of the maximiser. There are a number of abstract and high-dimensional versions of this methods in combination with large-deviation theory, e.g., [Bol86]. ◇

We now apply this idea to the study of the moments of the total mass of the PAM, to derive more precise asymptotics and to gain more insight in the critical behaviour of the path in the Feynman-Kac formula (2.2) or in the behaviour of the potential. More precisely, as we already mentioned in Remarks 3.8 and 3.14, one conceives the moments of $U(t)$ as an exponential moment of some functional whose minimiser and behaviour close to the minimiser is smooth, and derives a high-dimensional variant of the above idea. Since the moments of $U(t)$ can be seen as exponential moments in a two-fold way (for the path, see Sect. 3.3, and for the potential, see Sect. 3.2), there are also basically two ways to apply this idea. We concentrate here on the first one, the path-wise approach, as the potential-wise approach has been sufficiently commented on already in Remark 3.14 (annealed case in class (AB)) and Sect. 6.2 (quenched case in class (DE)); there are no more such results in the literature, to the best of our knowledge.

The path-wise version is usually carried out by a *Girsanov transformation*. It is based on the characteristic variational formula in (3.32). There are two analytic prerequisities that have to be ensured:

(i) the formula in (3.32) has (up to spatial shifts) precisely one minimiser, and
(ii) it is stable.

We reported in Sect. 3.4 on the validity of Property (i) in the respective cases.

In (ii), *stability* is meant in the sense that any sequence of admissible functions such that their functional values converge towards the minimum has a subsequence that converges towards the minimiser, up to spatial shifts. To be useful, this convergence must be valid in the topology of the LDP for the rescaled local times. Proving such a statement is by far not trivial and must be done on a case-by-case basis. Stability is known to hold for the problem (3.53) with $\gamma = 0$, which arises for Brownian motion among Poisson obstacles, i.e., for the difference of the Lebesgue measure of a set and its Dirichlet eigenvalue, see [Oss79] for $d = 2$ and [Hal92] for $d \geq 3$. For the variational problem that arises from the double-exponential distribution, (3.45), stability was proved in [GärHol99].

A deeper analysis of the PAM in the spirit of the Laplace method has been carried out in the discrete-space setting for range problem (see Example 1.10) in $d \leq 2$ [Bol94] and for the double-exponential distribution in any dimension [GärHol99], and in the continuous-space setting for the Brownian motion among Poisson obstacles in $d = 2$ [Szn91] and in $d \geq 3$ in [Pov99]. For both [Bol94] and [Szn91], the stability in $d \geq 3$ proved in [Hal92] came too late, but [Pov99] made up for this in $d \geq 3$, at least partly. The main results in [Szn91] and [Pov99] are sometimes called the *Brownian confinement property*, see Remark 3.14. See also

[Fuk08] for an extension of the confinement property to a mixture of 'hard' and 'soft' obstacles in $d \geq 2$ and a large-deviation principle for the rescaled Brownian endpoint with explicit identification of the rate function.

7.1.1 *Precise Asymptotics in Case (DE): Spatial Correlations and Effective Shift*

Let us describe how to implement the strong version of the Laplace method in the analysis of the total mass of the solution to the PAM and what we can learn from this. We do this only in the discrete-space setting with the potential ξ doubly-exponentially distributed with parameter ρ; we follow [GärHol99]. Here, the PAM with constant initial condition $\equiv 1$ is considered, in which case we write v for the solution and have the Feynman-Kac formula $v(t,x) = \mathbb{E}_x[\exp\{\int_0^t \xi(X(s))\,\mathrm{d}s\}]$ for $x \in \mathbb{Z}^d$ and $t \geq 0$.

The main goal of [GärHol99] is to analyse more terms of the expansion for the moments of $v(t,x)$ such that the first term that depends on the space variable x is identified. The main result there is that, as $t \to \infty$, for any $x, y \in \mathbb{Z}^d$,

$$\langle v(t,x)v(t,y)\rangle \sim \mathrm{e}^{H(t)-\chi t+C_2(t)} \sum_{z\in\mathbb{Z}^d} g(x+z)g(y+z), \tag{7.2}$$

where (as always) $H(t)$ is the logarithmic moment generating function of $\xi(0)$ and χ the characteristic variational formula in (3.45), and $C_2(t)$ is a function of order $o(t)$, which does not depend on x nor on y. (Its identification was beyond the scope of the paper [GärHol99], but it is presumably much smaller than just $o(t)$.) Here, $g\colon \mathbb{Z}^d \to (0,\infty)$ is equal to the minimiser of the formula in (3.45). We are under the assumption that, up to spatial shifts, there is precisely one minimiser, which is proved to be true as soon as $\rho > 15.7$. See Remark 3.17 for some properties of g.

The information that we draw from (7.2) is that the only asymptotic sensitivity of $\langle v(t,x)v(t,y)\rangle$ sits in a term of size $O(1)$, and that this can explicitly be written in terms of the solution of the characteristic formula. In particular, the correlations $\langle v(t,x)v(t,y)\rangle/\langle v(t,x)^2\rangle$ converge towards some non-trivial term that is sensitive to $x-y$. Using that v is a superposition of shifted versions of u, one can heuristically derive (7.2) from the eigenvalue expansion in (6.2).

As we will see, the implementation of the Laplace method will lead also to other deep results, like the characterisation of the random shift of g^2 to which the center of mass of the local times is attracted, the *effective shift*. We will record this in Theorem 7.2 below.

We give a glimpse of the proof of (7.2), by showing how to make the third term in the asymptotics $\langle v(t,x)\rangle = \mathrm{e}^{H(t)-\chi t+o(t)}$ explicit. Certainly, by shift-invariance of the distribution of the potential, $\langle v(t,x)\rangle$ does not depend on x at all, but the same procedure is suitable for proving (7.2), as we also will make clear. For this, we turn to the implementation of the step analogous to (7.1). Fix $x \in \mathbb{Z}^d$. Analogously to (2.14), we have

$$\langle v(t,x)\rangle = \mathbb{E}_x\Big[\exp\Big\{\sum_{z\in\mathbb{Z}^d} H(\ell_t(z))\Big\}\Big], \qquad t>0.$$

We know from Remark 3.14 that $\frac{1}{t}\ell_t$ is close to some shift of g^2 under the measure with density $\exp\{\sum_{z\in\mathbb{Z}^d} H(\ell_t(z))\}$, recall the tube property in (3.40). For the double-exponential distribution (certainly, with $\alpha(t) = 1$) this will be implicitly proved by the following; this gives the proof that we announced in Remark 3.14. We want to keep control on the shift, hence, we use the neighbourhood

$$\mathcal{U} = \bigcup_{w\in\mathbb{Z}^d} U(\theta_w(g^2)),$$

where $U(\mu)$ is some neighbourhood of a probability measure μ on $\mathbb{Z}^d$, and θ_w is the shift-operator by w. Think of $U(\mu)$ as being so small that the sets $U(\theta_w(g^2))$ with $w \in \mathbb{Z}^d$ are mutually disjoint. Hence, we see that

$$\begin{aligned}\langle v(t,x)\rangle &\sim \sum_{w\in\mathbb{Z}^d} \mathbb{E}_x\Big[\exp\Big\{\sum_{z\in\mathbb{Z}^d} H(\ell_t(z))\Big\}\mathbb{1}\{\tfrac{1}{t}\ell_t \in U(\theta_w g^2)\}\Big]\\ &= \sum_{w\in\mathbb{Z}^d} \mathbb{E}_{x-w}\Big[\exp\Big\{\sum_{z\in\mathbb{Z}^d} H(\ell_t(z))\Big\}\mathbb{1}\{\tfrac{1}{t}\ell_t \in U(g^2)\}\Big].\end{aligned} \tag{7.3}$$

Now we come to the Girsanov transformation that we announced. The important point is that the operator

$$G_g f(x) = \sum_{z\in\mathbb{Z}^d : z\sim x} \frac{g(z)}{g(x)}\big(f(z)-f(x)\big), \qquad x\in\mathbb{Z}^d, f\in\ell^2(\mathbb{Z}^d), \tag{7.4}$$

generates a random walk $(X_t)_{t\in[0,\infty)}$ on $\mathbb{Z}^d$ that is reversible with invariant measure equal to g^2, the minimiser in (3.45). The operator is symmetric on the space $\ell^2(\mathbb{Z}^d)$, equipped with the probability weight g^2. Since g decays quickly to zero far out, the transformed random walk has strong recurrence properties, and therefore we can expect that the normalised local times, $\frac{1}{t}\ell_t$, are close to g^2 for large t. We can give a formula for the density of the transformed Markov chain with respect to the simple random walk: For $t > 0$ and any event A of paths $[0,t] \to \mathbb{Z}^d$, we have, for any $x, z \in \mathbb{Z}^d$,

$$\mathbb{P}_x^{(g)}\big((X_s)_{s\in[0,t]} \in A, X_t = z\big) = \mathrm{e}^{\chi t}\mathbb{E}_x\Big[\exp\Big\{\int_0^t \rho \log g^2(X_s)\,\mathrm{d}s\Big\}\mathbb{1}\{(X_s)_{s\in[0,t]} \in A\}\mathbb{1}\{X_t = z\}\Big]\frac{g(x)}{g(z)}. \tag{7.5}$$

This actually follows from some knowledge about the variational formula (3.45). Indeed, the exponential density term is in general given as $\exp\{\int_0^t \frac{-\Delta^{\mathrm{d}} g}{g}(X_s)\,\mathrm{d}s\}$. However, we recall from (3.46) the Euler-Lagrange equation $\frac{-\Delta^{\mathrm{d}} g}{g} = \chi + \rho \log g^2$, which implies that (7.5) holds. Hence, a rewrite of the Feynman-Kac formula in terms of the transformed Markov chain derives from (7.3)

$$\langle v(t,x)\rangle = \mathrm{e}^{H(t)-\chi t}\sum_{w\in\mathbb{Z}^d} g(x-w)\,\mathbb{E}_{x-w}^{(g)}\Big[\mathrm{e}^{F_t(\frac{1}{t}\ell_t)}\mathbb{1}\{\tfrac{1}{t}\ell_t \in U(g^2)\}\tfrac{1}{g(X_t)}\Big], \tag{7.6}$$

where

$$F_t(\mu) = \sum_{z\in\mathbb{Z}^d}\Big(H(t\mu(z)) - \mu(z)H(t) - t\mu(z)\rho\log g^2(z)\Big), \qquad \mu \in \mathcal{M}_1(\mathbb{Z}^d).$$

We have now driven the approximation of the Feynman-Kac formula one step further than we did previously in Sect. 3.3. Thanks to Assumption (H) with $\eta(t) = t$ and $\widehat{H}(y) = \rho y \log y$ in (3.26), we know that $F_t(\frac{1}{t}\ell_t) = o(t)$ as $t \to \infty$, since $\frac{1}{t}\ell_t \approx g^2$ under $\mathbb{P}_x^{(g)}$.

The big task is now to show that the dependence of the last expectation in (7.6) on $x - w$ vanishes in the limit $t \to \infty$, i.e.

$$\mathbb{E}_x^{(g)}\Big[\mathrm{e}^{F_t(\frac{1}{t}\ell_t)}\mathbb{1}\{\tfrac{1}{t}\ell_t \in U(g^2)\}\tfrac{1}{g(X_t)}\Big] \sim \mathrm{e}^{C_1(t)}, \qquad x \in \mathbb{Z}^d,$$

where $C_1(t) = o(t)$ does not depend on x. This is very plausible, as the random walk has strong recurrence properties and therefore the distribution of X_t quickly converges towards the invariant distribution g^2 and therefore quickly loses its memory. This makes it also plausible that the limit $t \to \infty$ may be interchanged with the sum on w in (7.6). The technical obstacle is to separate the influence of the starting site $x - w$ from the term $\mathrm{e}^{F_t(\frac{1}{t}\ell_t)}\mathbb{1}\{\frac{1}{t}\ell_t \in U(g^2)\}$ and this from the term $1/g(X_t)$. This is done via a separation of time intervals $[0,r]$ and $[r, t-r]$ and $[t-r, t]$ for some large r that does not depend on t.

We terminate here our brief description of the procedure carried out in [GärHol99]. We only would like to mention that a variant of the periodisation technique that we outlined in Sect. 4.3 has to be incorporated into the proof as well, which makes it quite cumbersome.

From the above, we also obtain an explicit convergence of both $\frac{1}{t}\ell_t$ and X_t under the annealed path measure Q_t with density $\frac{1}{\langle U(t)\rangle}\exp\{\sum_{z\in\mathbb{Z}^d} H(\ell_t(z))\}$ (see (2.16)):

Theorem 7.2 (Limiting distribution of the effective shift) *Let W be a $\mathbb{Z}^d$-valued random vector with density $g/\|g\|_1$. Then the distribution of the normalised local times, $\frac{1}{t}\ell_t$, under Q_t converges towards $\delta_{\theta_W(g^2)}$, and the one of X_t converges towards the W-shift of g^2, i.e.,*

$$\lim_{t\to\infty} Q_t(X_t = z) = \sum_{w\in\mathbb{Z}^d} \frac{g(w)}{\|g\|_1} g^2(w+z).$$

In other words, the typical path in the Feynman-Kac formula for the total mass $U(t)$ picks with probability $g(w)/\|g\|_1$ a site w and then builds up local times with the shape $g^2(w+\cdot)$. This assertion has never been explicitly stated nor proved in the literature, to the best of our knowledge, but it follows from the above outline with the technical work done in [GärHol99].

The changes for deriving (7.2) from the above are the following. We write $v(t,x)v(t,y)$ in terms of an expectation with respect to two independent random walks, starting at x and at y, and we write the exponential term in terms of a functional of the sum of the two normalized local times of the walks. Then an obvious extension of the transition density in (7.5) to the pair of walks and a proper change of the definition of F_t is used to derive a formula analogous to (7.6). The technicalities to be resolved are the same.

7.2 Correlated Potentials

In the entire book so far, we considered only random potentials ξ on $\mathbb{Z}^d$ with independent and identically distributed values in the sites. This makes the mathematical analysis and the determination of the potential distribution easier; our classification of asymptotic behaviours in Sect. 3.4 is entirely based on the asymptotics of the upper tails of the single-site distribution. However, the assumption of independence of the potential variables appears as a mathematical idealisation, and many interesting random potentials indeed have long-reaching correlations, in particular in the spatially continuous case, as we have seen in many examples. (In this view, one should perhaps conceive the i.i.d. case in $\mathbb{Z}^d$ as a subcase of the $\mathbb{R}^d$-case with positive finite correlation length.) But also in the spatially discrete case, it appears highly interesting to study the PAM with potentials that have strong long-range correlations, as there might arise some new effects that change or even suppress the effect of intermittency and replace it by some homogeneous behaviour, as the highest peaks might be washed out severely. In this section, we report on what has been done in this respect.

We consider stationary potentials (i.e., potentials whose distribution is not changed under spatial shifts) and want to work with a non-rigorous notion of a correlation length; we want to understand this just as a spatial scale that roughly indicates the smallest distance of independent potential values. A white-noise potential on $\mathbb{R}^d$ has correlation length zero, Poisson potentials of the form $\sum_{i\in\mathbb{N}}\varphi(\cdot - x_i)$ with $(x_i)_{i\in\mathbb{N}}$ a Poisson point process in $\mathbb{R}^d$, and Gaussian fields have correlation length equal to the diameter of the support of the cloud φ, respectively of the covariance function of the Gaussian field. In the case of an infinite correlation length, the main interest is in the dependence of the long-time behaviour of the solution on the decay behaviour of the correlation, i.e., of the could φ in the Poisson case and of the covariance function in the Gaussian case, to name two examples.

In Examples 7.3 and 7.4 we first report on what is known for correlated fields under some abstract conditions, and in Example 7.5 we show how large the correlation length of the Poisson field, respectively of the Gaussian field, can be chosen without changing the behaviour of the solution. However, in Example 7.6 we go beyond this threshold and encounter a new effect. Throughout this section, u is, in the spatially discrete case, the solution to the PAM in (1.1) and (1.2), and $U(t)$ is its total mass; in the spatially continuous case, u is the solution to (1.7) and $u(0,\cdot) = \delta_0(\cdot)$ with total mass $U(t)$.

Example 7.3 (Correlated shift-invariant potentials on $\mathbb{Z}^d$) To the best of our knowledge, [GärMol00] is the only paper that studies the PAM on $\mathbb{Z}^d$ with correlated potentials. The field $\xi = (\xi(z))_{z\in\mathbb{Z}^d}$ is assumed stationary (i.e., shift-invariant in distribution), and the existence of the limit

$$\mathcal{H}(\mu) = \lim_{t\to\infty} \frac{1}{t} \log \frac{\Big\langle \exp\Big\{t\sum_{z\in\mathbb{Z}^d} \mu(z)\xi(z)\Big\}\Big\rangle}{\langle e^{t\xi(0)}\rangle} \tag{7.7}$$

for any probability measure μ on $\mathbb{Z}^d$ with compact support is assumed. This is obviously in the spirit of the class (DE) of Remark 3.17; indeed, in the special case of an i.i.d. potential satisfying Assumption (H) (see (3.26)) with $\eta(t) = t$, the limit in (7.7) exists with $\mathcal{H}(\mu) = \sum_{z\in\mathbb{Z}^d} \widehat{H}(\mu(z))$. Under the assumption in (7.7), the first two terms of the logarithmic asymptotics of the moments of the total mass are derived in [GärMol00] in terms of a characteristic variational formula. A heuristic derivation along the lines of the heuristics in Sect. 3.3 is as follows:

$$\begin{aligned}\langle U(t)\rangle &= \mathbb{E}_0\Big\langle \exp\Big\{t\sum_{z\in\mathbb{Z}^d} \xi(z)\tfrac{1}{t}\ell_t(z)\Big\}\Big\rangle \approx e^{H(t)}\mathbb{E}_0\Big[e^{t\mathcal{H}(\frac{1}{t}\ell_t)}\Big] \\ &\approx e^{H(t)}\exp\Big\{-t\inf_{\mu\in\mathcal{M}_1^{(c)}(\mathbb{Z}^d)}\big(I(\mu)-\mathcal{H}(\mu)\big)\Big\}\,e^{o(t)},\end{aligned} \tag{7.8}$$

where $\mathcal{M}_1^{(c)}(\mathbb{Z}^d)$ denotes the set of probability measures on $\mathbb{Z}^d$ with compact support, and we recall that $H(t) = \log\langle e^{t\xi(0)}\rangle$ is the logarithm of the moment generating function of $\xi(0)$, and $I(\mu) = \sum_{x,y\in\mathbb{Z}^d: x\sim y}(\sqrt{\mu(x)} - \sqrt{\mu(y)})^2$ is the (infinite-space version of) the large-deviation rate function for the normalized local times of the simple random walk, see Lemma 4.1. The above heuristics can be turned into a proof, since the characteristic variational formula possesses minimisers, and the diameter of the annealed intermittent island does not diverge as $t \to \infty$, like in the case (DE).

It would be of quite some interest to work out the details of the analogous situation in which also a spatial rescaling is involved, like in Assumption (H) for $\eta(t)$ that is not asymptotic to t. This class of correlated potentials could be designed to contain also some interesting Gaussian potentials with long-range correlations and more examples. ◇

Example 7.4 (General correlated shift-invariant potentials on $\mathbb{R}^d$) To the best of our knowledge, the only work on the PAM on $\mathbb{R}^d$ with general correlated potentials is [GärKön00]; it is the $\mathbb{R}^d$-variant of the work [GärMol00], see Example 7.3. Indeed, the main assumption on the potential ξ in [GärKön00] is the existence and non-triviality of the limit

$$\mathcal{H}(\mu) = \lim_{t\to\infty} \frac{1}{\beta(t)} \log\left\langle e^{\beta(t)\langle\mu,\xi_t\rangle}\right\rangle, \qquad \mu \in \mathcal{M}_1^{(c)}(\mathbb{R}^d), \tag{7.9}$$

where $\beta(t) = t\alpha(t)^{-2}$, and $\alpha(t)$ is some spatial scale in $(0,\infty)$, which may vanish or explode or stay constant as $t \to \infty$. The rescaled version ξ_t of ξ is defined as

$$\xi_t(x) = \alpha(t)^2\Big(\xi(\alpha(t)x) - \frac{H(t)}{t}\Big)$$

with $H(t) = \log\langle e^{t\xi(0)}\rangle$ as usual; compare to (3.19). Via the exponential Chebyshev inequality and a Legendre transformation, the assumption in (7.9) is essentially equivalent to the large-deviation principle of (3.14); the required sense of (7.9) is uniform on the set of all measures μ that are supported in a given compact subset of $\mathbb{R}^d$. Applications of Jensen's and Hölder's inequalities show that $\mathcal{H}$ is non-positive and convex.

The main result of [GärKön00] is, under some mild technical assumption on the asymptotics of $H(t)$, the expansion

$$\langle U(t)\rangle = e^{H(t)} \exp\Big\{-\beta(t) \inf_{\mu\in\mathcal{M}_1^{(c)}(\mathbb{R}^d)} \big(\mathcal{S}(\mu) - \mathcal{H}(\mu)\big)\Big\} e^{o(\beta(t))}, \qquad t\to\infty, \tag{7.10}$$

where $\mathcal{S}(\mu)$ is the rate function of the LDP in Lemma 4.3 without normalisation, more precisely, we put $\mathcal{S}(\mu) = \|\nabla g\|_2^2$ if g^2 is the density of μ and $g \in H^1(\mathbb{R}^d)$, and equal to ∞ if not. A heuristic derivation of (7.10) is analogous to (7.8), along the procedures outlined in Sects. 3.2 and 3.3; we do not give it here. The variational formula in (7.10) is strictly positive and nontrivial if $\mathcal{H}$ is nontrivial (i.e., not equal to zero everywhere). It can be evaluated quite explicitly in the important special case that $\mathcal{H}(\mu)$ depends on the covariance of μ only. ◇

Example 7.5 (Gaussian fields and Poisson shot-noise fields) In Examples 5.13 and 5.15 we summarised results from [GärKönMol00] about the almost sure asymptotics of the total mass of the solution to the PAM in the two interesting cases of a Hölder-continuous Gaussian potential and a Poisson shot-noise potential (i.e., with high peaks, unlike in the obstacle case) on $\mathbb{R}^d$. Here, we would like to point out that both results were derived under quite weak assumptions on the correlation lengths of the potential. Indeed, for the Gaussian potential, the first two terms in the almost sure asymptotics were shown not to depend on the details of the covariance function B, as long as it satisfies

$$\int_{([-R,R]^d)^c} g^2(x)\,\mathrm{d}x = o\big((\log R)^{-2/3}\big), \qquad R \to \infty,$$

where B can be presented as $B(x) = \int_{\mathbb{R}^d} g(x-y)g(y)\,\mathrm{d}y$, and g is the Fourier transform of the square root of the spectral density of the Gaussian field. This condition is a somewhat abstract assumption on the correlation length of the Gaussian field. One of the first technical steps in the proof of the almost sure asymptotics is to cut down the correlation length of the potential to finite size, for a later application of the Borel-Cantelli lemma.

The same paper also studied the Poisson potential of the form $\sum_{i\in\mathbb{N}} \varphi(\cdot - x_i)$, with $(x_i)_{i\in\mathbb{N}}$ a standard Poisson point process in $\mathbb{R}^d$ and φ a non-negative sufficiently regular cloud, taking its strict maximum at zero with a local parabolic shape. The first two terms of the asymptotics of the total mass were shown not to depend on the details of φ, as long as it decays fast enough, in the sense that it satisfies

$$\max_{([-R,R]^d)^c} \varphi = o\big((\log R)^{-1}\big), \qquad R \to \infty,$$

as long as the integrability condition $\int_{\mathbb{R}^d} \max_{x\in[-1,1]^d} \varphi(x-y)\,\mathrm{d}y < \infty$ is satisfied. This shows that the second-order asymptotics are very stable against long-term correlations in these two cases. ◇

In the following example, however, a very large choice of the correlation length was proved to have a quite different effect on both terms of the asymptotics.

Example 7.6 (Brownian motion among Poisson obstacles with long range) In the spatially continuous case, one considers the potential $V(\cdot) = -\sum_i W(\cdot - x_i)$ with $(x_i)_{i\in\mathbb{N}}$ a standard Poisson point process with parameter $\nu \in (0,\infty)$ in $\mathbb{R}^d$ and W a non-negative potential, i.e., the trap case. Let us assume that the cloud decays like $W(x) \approx |x|^{-q}$ as $|x| \to \infty$, for some $q > 0$. It was already shown in [DonVar75] that the first two terms in the asymptotics of the moments of the total mass are independent on q (and indeed the same as for $\varphi = -\mathbb{1}_{B_1}$, see Sect. 3.5.1), as long as $q > d+2$, i.e., the decay is strong enough.

However, in the interesting case $d < q < d + 2$, the moment asymptotics are different, and they indicate a dissolution of the concentration property by some homogenised behaviour. Indeed, [Fuk11] shows that they are given, for any $p \in [0, \infty)$, as

$$\langle U(t)^p \rangle = \exp\Big\{ - a_1 (pt)^{d/q} + (a_2 + o(1))(pt)^{\frac{q+d-2}{2q}} \Big\}, \qquad t \to \infty, \tag{7.11}$$

where

$$a_1 = \nu \omega_d \Gamma\Big(\frac{q-d}{q}\Big) \qquad \text{and} \qquad a_2 = \Big(\frac{\kappa \nu q \sigma_d}{2} \Gamma\Big(\frac{2q-d+2}{q}\Big)\Big)^{1/2}, \tag{7.12}$$

and ω_d and σ_d, respectively, are the volume and the surface of the unit ball in $\mathbb{R}^d$, and we have replaced Δ by $\kappa \Delta$ with some diffusion constant $\kappa \in (0, \infty)$. The first term in (7.11) was derived in [Pas77]. It is asymptotically equivalent to $\langle \mathrm{e}^{tV(0)} \rangle = \mathrm{e}^{H(t)}$, like in all the other cases that we encountered before. Let us also mention that the critical case $q = d + 2$ was studied by [Oku81].

The second term in (7.11) was studied and interpreted in [Fuk11]. The number a_2 admits a representation in terms of the variational formula

$$a_2 = \inf_{\phi \in H^1(\mathbb{R}^d) : \|\phi\|_2 = 1} \Big(\kappa \|\nabla \phi\|_2^2 + \frac{a_2^2}{\kappa d} \int_{\mathbb{R}^d} |x|^2 \phi(x)^2 \, \mathrm{d}x \Big).$$

The main joint strategy of the motion and the potential to contribute optimally to the Feynman-Kac formula is informally described as follows. Unlike in the standard case $q < d + 2$, there is no sharp interface between the area that is free of obstacles and is therefore covered by the Brownian motion local times. Instead, the obstacle density gets only gradually thinner away from the origin, and the influence of the long tails of the cloud stretches practically over all the space, but gets sufficiently weak only in a very small neighbourhood of the origin. The potential assumes its minimum at the origin with $|V(0)| \sim a_1 \frac{d}{q} t^{-(q-d)/q}$ and assumes a parabolic shape $V(x) - V(0) \sim \frac{a_2^2}{\kappa d} t^{-(q-d+2)/q} |x|^2$ for $|x| = o(t^{1/q})$. The Brownian motion does not leave the centred ball with radius $o(t^{1/q})$. This directly explains both terms of the asymptotics in (7.11), on base of the Donsker-Varadhan-Gärtner LDP for the occupation times measures of the Brownian motion.

In [Fuk11], consequences for the second-order asymptotics of the Lifshitz tails and of the almost sure asymptotics of $U(t)$ are drawn as well, but we do not formulate them here; this follows the usual patterns, given (7.11). ◇

In view of the current fruitful developments in the analysis of extreme-value properties of the Gaussian free field and other log-correlated random fields, it appears interesting to study the PAM with such a potential, in the hope to find also here some transition from a concentrated to a homogeneous behaviour of the solution.

7.3 Weak Disorder and Accelerated Motion

We saw already in our first considerations in Sect. 1.2 that the potential ξ makes the solution $u(t,\cdot)$ to the PAM in (1.1) irregular, and the Laplace operator makes it smooth. However, as we saw when discussing intermittency at various places, the smoothening effect is not so strong that it would entirely resolve the irregularity; a strong localisation effect is still dominant, even though the local areas of the high peaks show some smooth structures. This is rather different from what we see in other models of random motions through random media, for example the much-studied model of a random walk in random environment or the random walk with random conductances, which has a strong tendency to homogenisation. That is, this kind of random walk, under quite general conditions on the distribution of the random environment, satisfies a central limit theorem, and therefore spreads out on the space scale $\sqrt{t}$, and does not clump together on small islands. Roughly, the random walk in random potential shows a localised behaviour, and the random walk in random environment shows a homogenous one.

In this section we study the transition in the PAM from localised to homogenised behaviour by modifying the Anderson Hamiltonian in (1.1) by either weakening the potential or speeding-up the diffusivity by way of multiplication with a strong factor, depending on time. Highly interesting new phenomena arise here.

To this end, we look at the operators

$$e^{t(\Delta^{\rm d}+\varepsilon_t\xi)} \qquad \text{with } \varepsilon_t\downarrow 0, \qquad \text{and} \qquad e^{t(\kappa_t\Delta^{\rm d}+\xi)} \qquad \text{with } \kappa_t\to\infty. \tag{7.13}$$

Certainly, the consideration of these two operators is mathematically equivalent via the relation $\varepsilon_{t\kappa_t}=1/\kappa_t$. The small factor ε_t tames down the influence of the potential and makes the disorder weak, and the large prefactor κ_t induces an acceleration of the random motion and makes the diffusion fast. Generally, it is expected (and has been proved in a number of cases) that scale functions ε_t respectively κ_t that are not too fast will not change the general picture in the asymptotics that we have for the standard case $\varepsilon_t\equiv 1=\kappa_t$, but only the scales. On the other hand, extremely fast choices will make the random potential so marginal that its influence vanishes or is summarized by some single diffusion constant in terms of central-limit type behaviour.

Naturally, the question arises whether there are further interesting regimes, in particular critical ones, between these two. There are a number of aspects under which this is interesting:

- What is a critical scale on which the random walker is so fast that he/she cannot spend much time in the highest peaks of the potential? What does he/she do instead?
- What is the critical scale on which the extremely high potential peaks do not attract the main flow of the mass? What happens instead?
- What is a critical scale of ε_t, respectively of κ_t, such that the two terms describing the logarithmic asymptotics of the moments are merged, and how does this work?

Having found the critical scale, one naturally asks also for the size and the structure of the relevant regions and for the mechanism that is behind the main contribution to the total mass of the PAM (moderate deviations? central limit theorem? what else?).

Like in the standard case $\varepsilon_t \equiv 1 = \kappa_t$, the asymptotics of the operators in (7.13) is closely connected with the upper-tail behaviour of the principal eigenvalue of the Anderson operator $\Delta^{\rm d} + \varepsilon\xi$ with small prefactor $\varepsilon \in (0, \infty)$ in front of the disorder in large ε-dependent boxes, which is an interesting object to study on its own. One can expect (and this is one of the fundamental questions here) that, for sufficiently small boxes, the corresponding (random) principal eigenfunction shows a *homogeneous behaviour*, i.e., stretches its mass homogeneously over the entire box, while for very large boxes, it shows a *localized behaviour*, i.e., concentrates its mass in some small islands, in the way that we know from the description of the almost sure behaviour of the PAM in Sect. 5. In the first case, the influence of the random potential ξ should come only in terms of its expected value and variance in terms of a central limit theorem; this has been carried out in [BisFukKön16] for boxes with diameter of order ε^{-2}. In the second case, it should come via an extreme-value analysis of the principal eigenvalue in small boxes. However, it is not clear what should happen for boxes of intermediate sizes: is the mass stretched thinner and thinner homogeneously, or does it develop a number of bumps?

Let us describe some explicit examples that have been handled in the literature.

7.3.1 Acceleration of Motion

In [Sch10], various interesting choices of the velocity function $(\kappa_t)_{t>0}$ for various choices of potential classes (introduced in Sect. 3.3) are considered, see also the summary [KönSch12]. In each of the cases considered, the moment asymptotics for the total mass are identified in terms of a characteristic variational formula analogous to (3.32). An interesting competition between the growth of κ_t and the upper tails of ξ arises: the faster κ_t grows, the stronger the flattening effect of the diffusion term is. As usual, it is supposed that Assumption (H) holds (see (3.26)), i.e., regularity of the logarithmic moment generating function H at infinity.

Reference [Sch10] identifies two critical scales for κ_t. A lower critical scale (different from the one that we discussed above) is identified that marks the threshold between unboundedly growing intermittent islands and concentration in just one site. This scale depends on the upper tails of ξ and is equal to $\eta(t)/t$ in (3.26). Precisely at the critical scale $\kappa_t \asymp \eta(t)/t$, we have a discrete picture, i.e., the relevant islands have a non-trivial, discrete shape, like in the class (DE) in the standard case. Interestingly, for upper tails of ξ in the class (B), on this critical scale, the characteristic variational formula is equal to the discrete version of the formula for χ in the case (B), i.e., with $\mathbb{R}^d$ replaced by $\mathbb{Z}^d$.

Now, assuming that $\kappa_t \gg \eta(t)/t$, we now come to the second critical scale for κ_t, the one that is asked for at the beginning of this section, i.e., the one that marks the threshold between extreme values of the local principal eigenvalues and moderately large ones. This transition is also reflected by the fact that for slower functions κ the asymptotics are described in terms of just the upper tails of $\xi(0)$ like in the standard case, and for faster ones, the entire distribution of $\xi(0)$ enters the description. This critical scale is characterised by the fact that the local times per site stays bounded, the path covers a region of radius $\asymp t^{1/d}$ (i.e., of volume $\asymp t$), the term $\int_0^t \xi(X_s)\,\mathrm{d}s$ is essentially a sum over $O(t)$ i.i.d. random variables. Hence, a kind of moderate-deviation mechanism for the sum of about t potential values is combined with a large-deviation principle for the rescaled local times on a box of radius $\asymp t^{1/d}$, and both run on the exponential scale t.

Let us formulate the main result for κ_t being on the critical scale, see [KönSch12, Theorem 3.2]. We assume that $\langle \xi(0)\rangle = 0$ and write $U^{(\kappa_t)}$ for the total mass of the PAM with additional prefactor κ_t in front of the Laplace operator. Fix $\theta \in (0,\infty)$ such that $\kappa_t t^{-2/d} \to \frac{1}{\theta}$, then

$$\big\langle U^{(\kappa_t)}(t)\big\rangle = \exp\Big\{-\frac{t}{\theta}(\chi_H(\theta)+o(1))\Big\}, \tag{7.14}$$

where

$$\chi_H(\theta) = \inf_{g\in H^1(\mathbb{R}):\|g\|_2=1}\Big(\|\nabla g\|_2^2 - \theta\int H\circ g^2\Big), \tag{7.15}$$

where we wrote $\int H\circ g^2$ for short for $\int_{\mathbb{R}^d} H(g^2(x))\,\mathrm{d}x$. An elementary substitution of g with $g(\beta\,\cdot)\beta^{d/2}$ shows that $\frac{1}{\theta}\chi_H(\theta)$ is equal to $\theta^{-d/(d+2)}\chi_{H(\cdot\theta)}(1)$, a remark that helps understanding the result in (7.14) when following the heuristics in Sect. 3.3 (which we are not doing here) and helps also comparing to the main result (7.17) of Sect. 7.3.3. Note that all the variational formulas called $\chi_\circ$ in Sect. 3.4 are versions of $\chi_H(\theta)$ with special choices of H and θ.

This critical phase has not been deeper analysed, nor the (conjecturally, homogenised) phase where $\kappa_t \gg \eta(t)/t$. Analogously to the example in Sect. 7.3.3, some interesting phase transition(s) are to be expected in the behaviour of the minimisers of $\chi_H(\theta)$ in the parameter θ. More precisely, it is conjectured that, for θ sufficiently small, the infimum in (7.15) is only asymptotically attained at flatter and flatter functions g that approach the zero function, but for all larger θ, the minimum should be attained. Neither an analysis of the almost-sure behaviour of $U^{(\kappa_t)}(t)$ has yet been carried out.

7.3.2 Scaled Gaussian Potential

Using a combination of some results on stretched exponential moments of (renormalised) self-intersection local times of random walks on $\mathbb{Z}^2$, which are also interesting on their own, one finds another explicit example that lives on the critical scale in the terminology of Sect. 7.3.1. Let $d = 2$ and $\xi = (\xi(z))_{z\in\mathbb{Z}^d}$ be a collection of i.i.d. Gaussian random variables with mean zero and variance $\sigma^2 \in (0,\infty)$. Then the logarithmic moment generating function is $H(t) = \sigma^2 t^2/2$, hence we take the scale function $\eta(t)$ of (3.26) equal to t^2. We consider the 'accelerated' Laplace operator $\kappa_t\Delta^{\mathrm{d}}$ with $\kappa_t = t^{2/d} = t$ and keep the notation $U^{(t)}$ for the total mass of the solution. This means that we do not have $\kappa_t \gg \eta(t)/t$, i.e., the classification outlined in Sect. 7.3.1 does not apply; actually the two critical scales coincide for this distribution. Nevertheless, we are able in this special situation to find the large-t asymptotics of the expected total mass, and we see an interesting non-trivial phase transition.

Indeed, a simple rescaling and (2.14) show that

$$\langle U^{(t)}(t)\rangle = \Big\langle \mathbb{E}_0\big[\mathrm{e}^{\frac{1}{t}\int_0^{t^2}\xi(X_s)\,\mathrm{d}s}\big]\Big\rangle = \mathbb{E}_0\Big[\mathrm{e}^{\frac{1}{2}\sigma^2 t^{-2} S_{t^2}}\Big], \tag{7.16}$$

where $S_s = \sum_{z\in\mathbb{Z}^d}\ell_s(z)^2 = \int_0^s \mathrm{d}r\int_0^s \mathrm{d}\tilde r\,\mathbb{1}\{X_r = X_{\tilde r}\}$ denotes the self-intersection local time of the simple random walk $(X_t)_{t\in[0,\infty)}$; see also Example 7.9.

Luckily, the asymptotics of the last term in (7.16) can be found in the literature both for small and for large σ, at least in discrete time instead of continuous time. Indeed, combining Theorem 2.5 and Lemma 1.4 of [BrySla95] implies that, for σ^2 small enough, the last term is upper bounded against $Ct^{2\sigma^2/\pi}$ for all large t. (An inspection of the proof shows that one could show even asymptotic equivalence with this term, where the constant C is equal to the $\sigma^2/2$-th exponential moment of the renormalised self-intersection local time of the Brownian motion; see also [Che10, Sections 5.4 and 8.2].) Furthermore, the proof of [BrySla95, Theorem 2.5] also suggests that the underlying random walk (i.e., the one that gives the main contribution to the last term in (7.16)) is diffusive, i.e., runs on the scale $\sqrt{t}$, as our heuristics in Sect. 7.3.1 explain. These assertions are proved there only for discrete-time random walks, but we have no doubt that some similar statement should be true also for continuous-time walks.

In contrast, for sufficiently large σ^2, [BolSch97] (see Example 7.11) shows that the right-hand side of (7.16) does not run on a polynomial, but the exponential scale, and they derive its logarithmic asymptotics in terms of a variational formula in the spirit of the LDP in Lemma 4.1. Furthermore, they go much deeper into the rigorous description of the underlying random walk that gives the main contribution. In particular, a discrete picture arises, i.e., there is no spatial rescaling, and the random walk runs on a finite (in t) scale, it is self-attractive. See also [Bol02] for an extensive survey of [BolSch97]. It is unknown whether or not there is a gap between the two regimes and what happens in between them.

This result constitutes actually a special case of (7.14), including the suggested triviality of the variational formula (7.15) for small θ and the non-triviality for large θ. The determination of the threshold between the two is still open, even though [BasCheRos06] derived a good bound in terms of the *Gagliardo-Nirenberg formula*.

7.3.3 Brownian Motion in a Scaled Poisson Potential

In a series of papers [MerWüt01a, MerWüt01b, MerWüt02], Merkl and Wüthrich considered Brownian motion among soft Poisson obstacles (see Remark 1.15 and Sect. 3.5) with the potential $V(x) = -\beta\varepsilon_t \sum_{i\in\mathbb{N}} W(x - x_i)$, where $(x_i)_{i\in\mathbb{N}}$ is a standard Poisson point process in $\mathbb{R}^d$, $W: \mathbb{R}^d \to [0,\infty)$ is a bounded measurable and compactly supported cloud, and $\beta \in (0,\infty)$ is a parameter that gives rise to interesting phase transitions. We consider the moments of

$$U^{(\varepsilon_t)}(t) = \mathbb{E}_0\Big[\exp\Big\{-\beta\varepsilon_t \int_0^t \sum_{i\in\mathbb{N}} W(Z_s - x_i)\,\mathrm{d}s\Big\}\Big],$$

using a slight abuse of notation. Note that we are in the case (B) with $\gamma = 0$ in the terminology of Sect. 3.4. The scale function $(\varepsilon_t)_{t\in(0,\infty)}$ is chosen in critical way, i.e., in such a way that a new phenomenon arises. Recall from Sect. 3.5 that, in the standard case $\varepsilon_t \equiv 1$, the main contribution to the moments of $U^{(1)}(t)$ comes from Brownian paths running on the scale $t^{1/(d+2)}$, and the potential enters the leading asymptotic term, which is on the exponential scale $t^{d/(d+2)}$, only via the density parameter of the Poisson process (which we put equal to one here). In order to achieve a significant influence from the cloud W, one has to multiply it with $\varepsilon_t = t^{-2/(d+2)}$, which makes the term in the exponential running on the scale $t\varepsilon_t = t^{d/(d+2)}$, which is the scale on which the Brownian probability of not leaving a ball of radius $t^{1/(d+2)}$ runs. This choice of ε_t is also consistent with the choice $\kappa_t \asymp t^{2/d}$ in Sect. 7.3.1 and the above mentioned relation $\varepsilon_{t\kappa_t} = 1/\kappa_t$.

With this choice of ε_t, [MerWüt01a, Theorem 0.2(b)] says that

$$\langle U^{(\varepsilon_t)}(t)\rangle = \exp\Big\{-t^{d/(d+2)}\beta^{-2/(d+2)}(\chi_H(\beta) + o(1))\Big\}, \qquad t\to\infty, \tag{7.17}$$

where $\chi_H(\beta)$ is as in (7.15) with $H(t) = \mathrm{e}^{-t} - 1 = \log\langle \mathrm{e}^{tV(0)}\rangle$, the logarithmic moment generating function of the Poisson process.

Interestingly, for dimensions $d \geq 2$, there is a phase transition from small β to large β as to the structure of minimisers of $\chi_H(\beta)$; indeed, in [MerWüt01a, Theorem 0.3] it turns out that $\chi_H(\beta) = \beta$ for all sufficiently small positive β, but $\chi_H(\beta) > \beta$ for all other β. This can be interpreted by saying that the homogeneous phase (which is encountered in [MerWüt01a, Theorem 0.2(a)] when taking $\varepsilon_t \ll t^{-1/(d+2)}$) arises also on the critical scale $\varepsilon_t = t^{-2/(d+2)}$, if the prefactor β is small enough. However, a deeper analysis of this homogenised phase on path level or on potential level is still lacking. Let us also remark that for $\varepsilon_t \gg t^{-1/(d+2)}$, i.e., if the damping of the potential is not too strong, [MerWüt01a, Theorem 0.2(c)] proves asymptotics that are practically the same as in the standard case $\varepsilon_t = 1$, with a suitable adaptation of the scales.

In the follow-up papers [MerWüt01b, MerWüt02], the almost sure asymptotics of $U^{(\varepsilon_t)}(t)$ and large-deviation properties of the principal eigenvalue of $\frac{1}{2}\Delta + \varepsilon V$

in large, ε-dependent boxes are deduced from the result in (7.17) in a way that is analogous to the one that we described in Sect. 5.11. The correct choice for obtaining interesting new effects is $\varepsilon_t = (\log t)^{-2/d}$; similarly to the annealed setting, the arising variational formulas are proved to show 'homogenised' behaviour for $d \leq 3$ for small positive β and 'localised' behaviour for large β, but only 'localised' behaviour for $d \geq 4$. Note that the critical dimension is two for the annealed setting and four for the quenched one.

Remark 7.7 (Shrinking traps) Another way to weaken the interaction of the Brownian motion with the Poisson traps is to take the intensity of the Poisson process as a t-dependent, vanishing term, $\nu_t \downarrow 0$, but keeping the cloud fixed, i.e., taking $W = \mathbb{1}_K$ for some compact set K. The critical scale (i.e., the scale at which new effects arise, like above) is the choice $\nu_t = ct^{-2/d}$ with some $c > 0$ in $d \geq 3$ and $\nu_t = \frac{1}{t}\log^2 t$ in $d = 2$. For this choice, in [BerBolHol05] the annealed large-t logarithmic asymptotics of the total mass of the solution to the PAM are derived on the scale $t^{1-\frac{2}{d}}$ in terms of explicit variational formulas that are related to the formula appearing in (7.17), i.e., as $\chi_H(\beta)$ in (7.15) with $H(t) = \mathrm{e}^{-t} - 1$. Furthermore, the asymptotics as $c \downarrow 0$ and as $c \to \infty$ are analysed, and three phases are identified. Another novelty of [BerBolHol05] is the choice of the shape K of the traps as random with a certain distribution. The methods are based on the large-deviation principle that the team developed in [BerBolHol01] for the analysis of the moderate deviations of the Wiener sausage, see Example 7.10. ◇

Remark 7.8 (Scaled Gibbsian point field) In another follow-up paper [Mer03], a scaled Gibbsian point field is considered as in Example 1.17; see also Remarks 3.22 and 5.12. The prefactor of the Gibbsian field is picked in the same way as in the above case of a Poisson point field, and the results are of the same richness. However, the arising variational formulas crucially depend on the thermodynamic pressure of the system and are therefore analytically quite different from the formulas in the above case, more precisely, they are generalisations of them. Physical properties like the existence of phase transitions for the underlying Gibbs process enter the picture. ◇

7.3.4 Scaled Renormalised Poisson Potential

Recall the renormalized Poisson trap potential $V(x) = -\int_{\mathbb{R}^d} W(y-x)(\omega(\mathrm{d}y) - \mathrm{d}y)$ that we discussed in Remark 2.6, where $\omega = \sum_i \delta_{x_i}$ is a Poisson point process with intensity $\nu \in (0,\infty)$, and $W(x) = C|x|^{-q}$ for some $q \in (d/2, d)$. Recall that [CheKul12] proved that the solution to the continuous PAM in (1.7) exists and admits the Feynman-Kac formula, and the first moment of the solution is finite.

Now we add a factor of ε_t in front of the potential, i.e., we replace V by $\varepsilon_t V$, and we write $U^{(\varepsilon_t)}$ for the total mass of the solution. We want to understand the large-t behaviour of the first moment of the total mass at time t. Recall the formula (2.6), which we record here once more:

$$\langle U^{(\varepsilon_t)}(t)\rangle = \Big\langle \mathbb{E}_0\Big[\exp\Big\{\varepsilon_t \int_0^t V(Z_s)\,\mathrm{d}s\Big\}\Big]\Big\rangle = \mathbb{E}_0\Big[\exp\Big\{\nu \int_{\mathbb{R}^d} F(\varepsilon_t w(t,x))\,\mathrm{d}x\Big\}\Big],$$

where $F(x) = \mathrm{e}^{W(x)} - 1 + W(x)$ and $w(t,x) = \int_0^t W(Z_s - x)\,\mathrm{d}s$.

In [CheKul11], it turned out that the critical scale for ε_t is equal to $t^{-1-q/(d+2)}$ rather than $t^{-2/(d+2)}$. More precisely, we have

$$\log\langle U^{(\varepsilon_t)}(t)\rangle \sim \begin{cases} (t\varepsilon_t)^{d/q}\nu \int_{\mathbb{R}^d} F(\theta|x|^{-p})\,\mathrm{d}x & \text{if } \varepsilon_t \ll t^{-1-q/(d+2)}, \\ -t^{d/(d+2)}\chi_{r,C}, & \text{if } \varepsilon_t \sim rt^{-1-q/(d+2)}, \\ -\varepsilon_t^{4/(d+2-2q)} t^{(d+4-2q)/(d+2-2q)}\chi_C, & \text{if } \varepsilon_t \gg t^{-1-q/(d+2)}. \end{cases} \tag{7.18}$$

where

$$\chi_{r,C} = \inf_{g\in H^1(\mathbb{R}^d):\|g\|_2=1} \Big(\|\nabla g\|_2^2 - \nu \int_{\mathbb{R}^d} F\Big(Cr \int_{\mathbb{R}^d} \frac{g^2(y)}{|y-x|^q}\,\mathrm{d}y\Big)\,\mathrm{d}x\Big), \tag{7.19}$$

and χ_C is the same as $\chi_{1,C}$ with F replaced by the function $x \mapsto x^2/2$. Applications of these asymptotics to the Lifshitz tails of the underlying random Schrödinger operator are provided in [CheKul11] as well, along the relation outlined in Sect. 2.2.6.

Note that χ_C formally gives the main term in the small-r asymptotics of $\chi_{r,C}$, as $F(r) \sim r^2/2$ for $r \to 0$. The result in the last regime needs the assumption that $q < (d+2)/2$, which is also the assumption under which $w(t,x)$ is square-integrable, and this makes possible the application of a large-deviation result for Brownian motion in a Brownian sheet. It is instructive to compare to the asymptotics in the usual Poisson obstacle potential; see Sect. 3.5.1 and in particular (3.58) and (3.59), jointly with (3.53).

See [CheXio15] for an extension to the case where the Poisson process ω is replaced by the time-dependent potential $(\omega_s)_{s\in[0,\infty)}$ that consists of a family of independent Brownian motions such that ω_s is a Poisson point process in $\mathbb{R}^d$ for any s (however, with ε_t replaced by one). Here again a number of different regimes and effects appear that are of similar nature.

7.4 Upper Deviations of Random Walk in Random Scenery

As we mentioned in Remark 2.8, the term $\int_0^t \xi(X_s)\,\mathrm{d}s$ in the exponent of the Feynman-Kac formula in (2.2) is sometimes called the *random walk in random scenery (RWRSc)*. In the study of the PAM, we are most interested in the behaviour of its exponential moments. However, recently a number of researchers got interested in the question about the *upper deviations* of the RWRSc on various scales, i.e., in results of the type

$$\log \mathrm{Prob} \otimes \mathbb{P}\Big(\int_0^t \xi(X_s)\, \mathrm{d}s > \lambda a_t \Big) \sim -b_t I(\lambda), \qquad t \to \infty, \text{ for } \lambda \in (0,\infty), \tag{7.20}$$

where a_t and b_t are scale functions that tend to ∞ as $t \to \infty$. The motivation stems from a general interest in the behaviour of random motions in a random potential and the pull that the richness of phenomena obtained in the study of these questions exerts, furthermore from the mathematical challenges that the proofs put. In Remark 3.11, we explained that such a result is essentially equivalent to asymptotics for the exponential moments of the form

$$\log \Big\langle \mathbb{E}\Big[\exp\Big\{ \beta \frac{b_t}{a_t} \int_0^t \xi(X_s)\, \mathrm{d}s \Big\} \Big] \Big\rangle \sim b_t \sup_{\lambda \in (0,\infty)} \big[\lambda\beta - I(\lambda) \big], \qquad t \to \infty, \text{ for } \beta > 0, \tag{7.21}$$

i.e., to moment asymptotics of the total mass of the PAM with scaled potential, as we discussed in Sect. 7.3. Hence, (7.20) may a priori prove very useful for understanding the solution to the PAM at time t with potential $\frac{b_t}{a_t}\xi$. However, it may happen that the supremum on the right-hand side of (7.21) is trivial (i.e., constantly equal to zero or to infinity), and hence it must be checked on a case-by-case basis whether (7.20) is useful for the study of the PAM or not.

Nevertheless, the way of thinking about and the proof techniques for deriving results like in (7.20) are very close to the methods that we encountered, notably in Chap. 4, and both topics benefitted from each other over the past two decades. At this place, we do not want to go into details, but only mention that [AssCas07] provides a survey on results of the type (7.20). They are sometimes also called *moderate-deviations results*, depending on the choices of a_t and b_t.

Note that, for most of the potential distributions that we considered and for most of the prefactors in front of the potential ξ, the moment asymptotics of the total mass show two quite different terms, not just one like in (7.21). Hence, the above duality can be helpful in this form only in cases described in Sect. 7.3, or they have to be modified accordingly. This is the reason that most of the results of the form (7.20) are not helpful for the understanding of the PAM.

7.5 Upper Deviations of Self-Attractive Functionals of Local Times

Carrying out the expectation over the random potential ξ, we saw in Sect. 2.1.4 that the expectation of the total mass of the PAM is equal to the exponential moment of $\sum_{z\in\mathbb{Z}^d} H(\ell_t(z))$, where H is the logarithm of the moment generating function of $\xi(0)$. Analogously to the RWRSc in Sect. 7.4, moderate-deviation results for the upper tails of this functional are interesting on their own and have been intensively studied in the last two decades. Analogously to the relation between (7.20) and (7.21), such results may also prove helpful for the study of the moments of the PAM, with possibly rescaled versions of H, i.e., with t-depending prefactors in front of the functional. For such an application, one has to check, on a case-by-case basis, whether the rescaling of H comes from the PAM at time t with a potential $\varepsilon_t\xi$ for some scale function ε_t.

However, also without reference to the PAM, upper deviations of the functional $\sum_{z\in\mathbb{Z}^d} H(\ell_t(z))$ are interesting, and we now briefly mention some works on this aspect.

Example 7.9 (Self-intersections) One of the most-studied example is the *self-intersection local time (SILT)* of the random walk, where we take $p \in \mathbb{N}$ and $H(l) = l^p$. Then $\sum_{z\in\mathbb{Z}^d} H(\ell_t(z)) = \|\ell_t\|_p^p$ is the p-th power of the p-norm of the vector of local times. This is equal to the p-fold SILT (or just SILT for $p = 2$), i.e., to the total mass of time vectors at which the path has the same location, as we have

$$\begin{aligned}\|\ell_t\|_p^p &= \sum_{z\in\mathbb{Z}^d} \ell_t(z)^p = \sum_{z\in\mathbb{Z}^d} \int_0^t \mathrm{d}t_1 \dots \int_0^t \mathrm{d}t_p \prod_{i=1}^p \mathbb{1}\{X(t_i) = z\}\\ &= \int_0^t \mathrm{d}t_1 \dots \int_0^t \mathrm{d}t_p \, \mathbb{1}\{X(t_1) = \dots = X(t_p)\}.\end{aligned}$$

Certainly, one generalizes the study of $\|\ell_t\|_p^p$ to any $p \in [0,\infty)$, with the understanding that the case $p = 0$ is the range case since $\sum_z \ell_t(z)^0 = \sum_z \mathbb{1}\{\ell_t(z) > 0\} = |X([0,t])|$, and $p = 1$ is trivial since $\|\ell_t\|_1 = t$. Note that $p > 1$ is the self-attractive case (here H is convex) and $p \in (0,1)$ is the self-repellent case (H is concave), if *upper* deviations are considered. The logarithmic moment generating function of the Gaussian, and more generally, of the Weibull distribution with power parameter p, is asymptotically equal to t^p for large t, hence upper tails, respectively exponential moments, of the SILT are interesting for the PAM. (This is particularly explicit in Sect. 7.3.2, where we consider a rescaling of an i.i.d. Gaussian potential ξ in $d = 2$.) Furthermore, upper tails of the SILT (also for discrete-time random walks and, struggling with some delicate issues of renormalisation, also for Brownian motion) are of high interest for some applications in physics and because of the mathematical challenges that their study offers. Like for RWRSc (see Sect. 7.4), the study of the upper tails of the SILT and of the PAM mutually benefitted from each other in the last decades. The main reference is [Che10], but see also the short survey [Kön10] on some heuristics and some proof methods. ◇

Example 7.10 (Volume of Wiener sausage) Let us shed a light on an important special case of Example 7.9, having some connections to the PAM with scaled Poisson field that we described in Sect. 7.3.3. This is the Brownian version of the range case, i.e., the class where $H(t) = -\mathbb{1}_{(0,\infty)}(t)$. For a Brownian motion $(Z_s)_{s\in[0,\infty)}$, one looks at the volume of the ‘sausage’ (sometimes called the *Wiener* *sausage*), which is the random variable $S_a(t) = |\bigcup_{s\in[0,t]} K_a(Z_s)|$, where $K_a(x)$ is the ball of radius $a > 0$ centred at $x \in \mathbb{R}^d$. We saw in Sect. 3.5.1 that the expectation of the total mass of the solution to the PAM with a Poisson field of obstacles is equal to the negative exponential rate of $S_a(t)$. (We recall that this is the continuous analogue of the PAM on $\mathbb{Z}^d$ with Bernoulli-distributed potential ξ.) Hence, it appears interesting to look at the lower deviations for $S_a(t)$ on various scales $b_t > 0$, i.e., at the logarithmic asymptotics of $\mathbb{P}(S_a(t) \leq b_t)$. Indeed, there are a number of works on that question, see e.g., [BolHol94].

Here we would like to point at the work [BerBolHol01], as it works precisely on the interesting scale that is also considered in the study of the corresponding PAM case, i.e., for critically rescaled Poisson obstacle field, see Sect. 7.3.3.

Indeed, in [BerBolHol01] a moderate-deviations principle for $S_a(t)$ was derived featuring a variational formula that displays the same objects as in (7.15) with $H(t) = \mathrm{e}^{-t} - 1$. Analogously to (7.20) and (7.21), it stands in a Legendre-transform relation to that formula, which is the one that appeared in the result (7.17). Interesting phase transitions appear that partially relate to each other. Reference [BerBolHol01] stresses the description of the path behaviour and the large deviations for the occupation measures of the motions, while [MerWüt01a] mostly talk about the rescaled random potential. The self-attraction is so weak that the Wiener sausage leaves many holes and nowhere clumps together with high intensity; this is the path-picture analogous to the potential-picture described in [MerWüt01a]. This is in one sense a boundary case of the situation in the rescaled LDP in Lemma 4.3 (translated to the Brownian setting).

Let us remark that the approach used and further developed in [BerBolHol01] is different from the one of most of the papers that we cite in this book, which all use, in one or another way, the Donsker-Varadhan-Gärtner LDP of Sect. 4.2.1. Rather, the path under consideration is decomposed into time lags, and the interaction is conceived as a sum of a functional of all the time lags, after applying a sophisticated estimate that is able to handle the interaction between any two distant ones. Finally, an LDP for the empirical pair measures of the sequence of subpaths is employed. The authors stress that this method also can produce a proof for the LDPs of Sect. 4.2.1. The – even more interesting – case of the intersection of two independent Wiener sausages was handled by the same team in [BerBolHol04], using the same approach. One more application of that approach – again by the same team – was carried out in [BerBolHol05] for a weak-interacting version of the Brownian motion among Poisson traps, where the intensity of the Poisson process depends on t and shrinks to zero on a critical scale; see Remark 7.7. ◇

7.6 Self-Attractive Path Measures

We mentioned in Sect. 2.1.5 that the transformed path measure Q_t with density $\frac{1}{\langle U(t)\rangle} \exp\{\sum_{z\in\mathbb{Z}^d} H(\ell_t(z))\}$ (with $\ell_t(z) = \int_0^t \delta_{X(s)}(z)\,\mathrm{d}s$ the local times of the random walker) is self-attractive by convexity of H, the logarithm of the moment generating function of $\xi(0)$; see the argument in (4.5). The study of such measures attracted a number of researchers over the last two decades, also with H not coming from some random variable and without reference to the PAM, and also for H a concave function, where Q_t is self-repellent. We restrict ourselves to mentioning just two works in the self-attractive case.

Example 7.11 (Self-attractive measures with mean-field interaction) In [BolSch97], transformed path measures with density

$$\exp\Big\{\frac{1}{t}\sum_{z,y\in\mathbb{Z}^d}\ell_t(z)\ell_t(y)V(y-z)\Big\}$$

(properly normalised) were analysed in depth with techniques that are based on large-deviation theory, but go much beyond. Here $V\colon\mathbb{Z}^d\to[0,\infty)$ is a non-negative function with finite support. The interaction in this density is a self-attractive mean-field interaction and is in the spirit of the mean-field variant of the polaron model, see [DonVar83]. The main example is $V=\beta\delta_0$, such that the term in the exponential is equal to $\frac{1}{t}$ times the self-intersection local time of the random walk, see Sect. 7.3.2. For any such V, the path measure is self-attractive. The main assumption in [BolSch97] is that the self-attraction is sufficiently strong in some sense, i.e., that β is large enough, but sharpness of this condition is not analysed. (Recall our discussion of a particular case (critically rescaled Gaussian field in $d=2$) from Sect. 7.3.2, where we pointed out what happens for small β in the case $V=\beta\delta_0$.)

Then the main results of [BolSch97] are

- a variational formula for the large-t exponential rate of the partition function (the expectation of $\exp\{\frac{1}{t}\sum_{z,y\in\mathbb{Z}^d}\ell_t(z)\ell_t(y)V(y-z)\}$) on the scale t,
- the convergence of the distribution of the normalized local times $\frac{1}{t}\ell_t$ under the transformed path measure towards the minimiser of that variational formula, and
- the convergence of the endpoint of the path towards some closely related quantity.

On the technical side, this work is similar to the work done in [GärHol99] on the PAM in the case (DE), which we outlined in Sect. 7.1.1. See also [Bol02] for an extensive survey of [BolSch97].

Let us remark that a deeper analysis of the spatially continuous version, the mean-field polaron measure for the three-dimensional Brownian motion $(Z_s)_{s\in[0,\infty)}$ with interaction

$$\frac{1}{T}\int_0^T\int_0^T \mathrm{d}s\mathrm{d}t\frac{1}{|Z_s-Z_t|},$$

recently received new swing [KönMuk15, BolKönMuk15] by the introduction of the compactification method in [MukVar15], see Sect. 4.11. ◇

Example 7.12 (Self-attractive discrete-time random walks) References [IofVel12a] and [IofVel12b] consider a discrete-time random walk $(X_n)_{n\in\mathbb{N}_0}$ in $\mathbb{Z}^d$ with drift under the path measure with density $\exp\{\sum_{z\in\mathbb{Z}^d} H(\ell_n(z))\}$ (properly normalised), if n denotes the number of steps and $\ell_n(x) = \sum_{k=0}^n \mathbb{1}\{X_k = x\}$ is the local time. The interaction function H is taken self-attractive in the sense that $-H$ is subadditive ($H(n+m) \geq H(n) + H(m)$ for every $n, m \in \mathbb{N}$) and sublinear ($\lim_{n\to\infty} H(n)/n = 0$). The main attention is payed to the dependence of the limiting behaviour of the endpoint on the scale n as a function of the strength of the drift. More precisely, [IofVel12a] and [IofVel12b] derive LDPs and laws of large numbers and central limit theorems for X_n/n under the transformed path measure in the case that its expectation is bounded away from zero, i.e., that one has a ballistic behaviour. This issue is the object of Sect. 7.10 below, which is a quite big research topic that goes beyond the scope of the present text. ◇

7.7 Transition Between Quenched and Annealed Behaviour

Via the spatial ergodic theorem, the expectation of the total mass $U(t)$ of the solution to the PAM can be seen as an almost sure ergodic limit of mixtures of shifts of the solution. This is formulated as follows. Let $v\colon [0,\infty) \times \mathbb{Z}^d$ be the solution of (1.1) with the localised initial condition δ_0 replaced by the homogeneous condition $v(0,z) = 1$ for every z, see Remark 1.5. The corresponding Feynman-Kac formula reads $v(t,z) = \mathbb{E}_z[\mathrm{e}^{\int_0^t \xi(X_s)\,\mathrm{d}s}]$, as we remarked in Sect. 2.1.2. In particular, $U(t) = v(t,0)$. Now the spatial ergodic theorem gives, for fixed $t > 0$,

$$\langle U(t)\rangle = \langle v(t,0)\rangle = \lim_{B\to\mathbb{Z}^d} \frac{1}{|B|} \sum_{x\in B} v(t,x), \qquad \text{almost surely,} \tag{7.22}$$

where the limit is along centred boxes. This has quite some relevance for the experimental identification of $\langle U(t)\rangle$, since it would be rather cumbersome to realise many independent copies of the random potential ξ; this task is replaced by evaluating the Feynman-Kac formula at many values of the starting site of the random walk with the same realisation of the potential.

One would like to use (7.22) also for the experimental evaluation of the large-t asymptotics of $\langle U(t)\rangle$, and for this one must pick some t-dependent boxes instead of B. However, one must be careful with the choice of the size: too small box sizes do not suffice, as they do not give enough space to the system to equilibrate sufficiently fast, and the result would be as if one does not take the average of the box, i.e., the quenched behaviour described in Sect. 5. On the other hand, too large boxes do describe the annealed behaviour described in Sect. 3, but are too costly in terms of computing time. Hence, an interesting question is to determine what phenomena we will see for what scales of the box size.

This has been studied in [BenMolRam05] for the special case of Bernoulli traps (see Example 1.10), in [CraMol07] for the time-dependent Gaussian white-noise potential on $\mathbb{Z}^d$ (which we will discuss briefly in Sect. 8.3 below), in [BenMolRam07] for quite general random potentials as well as in [GärSch15] for potentials in the classes (DE) and (SP), see Sect. 3.4. In [BenMolRam05] and [BenMolRam07], weak laws of large numbers and central limit theorems are found, but in [GärSch15], convergence towards a stable limit law is derived. The main assertions are the following. In [BenMolRam05] and [BenMolRam07], under certain regularity assumptions similar to Assumption (H) in Sect. 3, for the scale function $J(t) = \log\langle U(t)\rangle$, two critical points $0 < \gamma_1 < \gamma_2 < \infty$ are found such that, using the notation $B_r = [-r, r]^d \cap \mathbb{Z}^d$,

$$\frac{1}{|B_{\gamma J(t)}|}\sum_{x\in B_{\gamma J(t)}} \frac{v(t,x)}{\langle v(t,0)\rangle} \stackrel{t\to\infty}{\to} \begin{cases} 1 & \text{if } \gamma > \gamma_1, \\ 0 & \text{if } \gamma < \gamma_1, \end{cases} \qquad \text{in probability,} \tag{7.23}$$

and

$$\frac{1}{|B_{\gamma J(t)}|}\sum_{x\in B_{\gamma J(t)}} \frac{v(t,x)-\langle v(t,0)\rangle}{\sqrt{\langle v(t,0)^2\rangle}} \stackrel{t\to\infty}{\Longrightarrow} \begin{cases} \mathcal{N} & \text{if } \gamma > \gamma_2, \\ 0 & \text{if } \gamma < \gamma_2, \end{cases} \tag{7.24}$$

where $\mathcal{N}$ denotes a standard Gaussian random variable. In contrast, in [GärSch15], under some regularity assumption that guarantees that ξ lies in class (DE) or (SP), for any $\alpha \in (0, 2)$, two scale function $L_\alpha(t)$ and $B_\alpha(t)$ are determined in dependence on the tails of $\xi(0)$, such that

$$\frac{1}{|B_{L_\alpha(t)}|}\sum_{x\in B_{L_\alpha(t)}} \frac{v(t,x)-A_\alpha(t)}{B_\alpha(t)} \stackrel{t\to\infty}{\Longrightarrow} \mathcal{F}_\alpha, \tag{7.25}$$

where $\mathcal{F}_\alpha$ is a stable distribution with parameter α, and $A_\alpha(t) = 0$ for $\alpha \in (0, 1)$, and $A_\alpha(t) = \langle v(t,0)\rangle$ for $\alpha \in (1,2)$ and $A_1(t) = \langle v(t,0)\mathbb{1}\{v(t,0) \le B_1(t)\}\rangle$ for $\alpha = 1$.

Let us also remark that in [BenBogMol05] the diffusive part is dropped, i.e., random variables of the form $\sum_{i=1}^{N_t} \mathrm{e}^{tX_i}$ with i.i.d. variables X_i and various choices of the scale function $t \mapsto N_t$ are considered. This can be seen as a preliminary study of the above, but has also its own interest as an extension of the *random energy model* (where one takes the X_i as standard normal variables), which is itself a toy model for spin systems. Indeed, the main idea of the proofs in [GärSch15] is a regular decomposition of the large box into smaller boxes B_i, in which the contribution to $v(t,x)$ is approximated by $\mathrm{e}^{t\lambda_i}$ with λ_i the local principal eigenvalue of $\Delta^{\mathrm{d}} + \xi$ in B_i; these eigenvalues are essentially i.i.d. Then one is in the situation of [BenBogMol05].

7.8 Other Diffusivities

So far, we considered the PAM with the diffusivity given only by the (discrete or continuous, respectively) Laplace operator, i.e., with a driving random motion that lies in the domain of attraction of the Brownian motion. In the characteristic variational formula describing the large-t asymptotics of $\langle U(t)\rangle$ (see, e.g., (3.32)), this type of diffusivity is seen in terms of an energy term $\|\nabla g\|_2^2$ coming from the Laplace operator (after possibly rescaling time and space in the spirit of Donsker's invariance principle).

However, instead of a motion taken from the Brownian universality class, it is certainly also highly interesting to study the case of random walks in the domain of attraction of Lévy processes, i.e., a random walk that can make very large steps. So far, the first rigorous, and currently only, work in that direction seems to be [MolZha12], where the Laplace operator Δ^{d} is replaced by the generator $\mathcal{L}f(z) = \sum_{x\in\mathbb{Z}^d} a(x)[f(x+z) - f(z)]$, where the weights $a(x)$ are supposed to satisfy $a(x) \approx |x|^{-2-\beta}$ as $|x| \to \infty$ for some $\beta \in (0,2)$. The potential ξ is supposed i.i.d. with Weibull tails, i.e., $\mathrm{Prob}(\xi(0) > r) = \exp\{-\frac{1}{\alpha}r^\alpha L(r)\}$ with some $\alpha \in (1,\infty)$, where L is a slowly varying function. One of the main results [MolZha12, Theorem 2.1] is that, in the case that $L(x) \equiv 1$, for any $p \in \mathbb{N}$, as $t \to \infty$,

$$t^{-\alpha'} \log\langle U(t)^p\rangle = \frac{p^{\alpha'}}{\alpha'}(1+o(1)), \qquad \text{where } \frac{1}{\alpha} + \frac{1}{\alpha'} = 1.$$

Furthermore, [MolZha12, Theorem 2.5] states that, almost surely,

$$\limsup_{t\to\infty} t^{-\alpha'} \log U(t) \le \frac{1}{\alpha'} \qquad \text{and} \qquad \liminf_{t\to\infty} t^{-\alpha'} \log U(t) \ge \frac{1}{\alpha'}\Big(\frac{d}{d+\alpha'-1}\Big)^{\alpha'/\alpha}. \tag{7.26}$$

The lower bound in (7.26) comes from the strategy of the random walk in the Feynman-Kac formula to wait at the origin for ≤ 1 time units and then to jump immediately to the site where the potential ξ attains its maximum within a centred box with diameter of order $t^{\alpha'}$. Further results specify the asymptotics for various choices of L; they differ from the above by some lower-order terms.

Hence, qualitatively, the asymptotics and their logistics follow the pattern that we outlined in Sects. 3 and 5 in the case (SP), but the proof is much simpler, as the path from the origin to the optimal island just makes one single jump. The gap between the almost sure lower and upper bound will give rise to future work; the authors conjecture that this is not only technical.

7.9 PAM in a Random Environment

Another interesting direction is the study of the PAM with the driving Laplace operator replaced by a random version of it, or, equivalently, with the underlying random walk replaced by a *random walk in random environment (RWRE)*. This makes the PAM as a model for random motions through random potential much more realistic, as the diffusion is now itself taken random, which represents impurities in the diffusive medium, hampering or accelerating locally the conductivity. There are numerous potential applications of such a additional randomness.

7.9.1 PAM with Random Conductances

One very natural way to introduce randomness in the diffusion is to replace Δ^{d} by the *randomised Laplace operator*,

$$\Delta^{\mathrm{d}}_{\omega} f(x) = \sum_{y\in\mathbb{Z}^d: y\sim x} \omega_{x,y}(f(y) - f(x)), \qquad f: \mathbb{Z}^d \to \mathbb{R},\ x \in \mathbb{Z}^d, \tag{7.27}$$

where $\omega = (\omega_{x,y})_{x,y\in\mathbb{Z}^d, x\sim y}$ is a random i.i.d. field of positive weights on the nearest-neighbour bonds of $\mathbb{Z}^d$. In order to obtain a symmetric operator $\Delta^{\mathrm{d}}_{\omega}$, one usually assumes that $\omega_{x,y} = \omega_{y,x}$ for any edge $\{x, y\}$, i.e., one attaches the weights to the undirected edges. One often also speaks of the *random conductance model*, since $\omega_{x,y}$ is interpreted as the conductance of the edge $\{x, y\}$. The operator $\Delta^{\mathrm{d}}_{\omega}$ generates the continuous-time random walk $(X_t)_{t\in[0,\infty)}$ in $\mathbb{Z}^d$, the *random walk among random conductances (RWRC)*. When located at y, it waits an exponential random time with parameter $\pi(y) = \sum_{z\in\mathbb{Z}^d: z\sim y} \omega_{y,z}$ (i.e., with expectation $1/\pi(y)$) and then jumps to a neighbouring site z' with probability $\omega_{y,z'}/\pi(y)$. The is studied a lot in the last decade like many other types of RWREs, with strong emphasis on the search for laws of large numbers or (functional) central limit theorems; see the survey [Bis11]. Here we are interested in the behaviour of the RWRC in boxes that execute some pressure on the path.

The PAM with RWRC as the underlying random motion has not yet been studied (however, note the work [ErhHolMai15b] on the case of a time-dependent potential with a RWRC with uniformly elliptic random conductances, see Remark 8.7), but important prerequisities have been derived: annealed large-deviation principles (LDPs) for the normalised local times of the RWRC in fixed boxes [KönSalWol12] and in time-depending, growing boxes [KönWol15]; see also the thesis [Wol13]. These LDPs are particularly interesting because of the assumption on the conductances that they are not uniformly elliptic (i.e., not bounded away from zero and from infinity), but can attain arbitrarily small values. This is precisely what creates an interesting interaction between the random walk and the random conductances, since small conductances help the random walk to lose much time within the box, i.e., to increase the probability of not leaving it. Putting assumptions on the lower tails of the conductances of the form

$$\log P(\omega_e \le \varepsilon) \sim -D\varepsilon^{-\eta}, \qquad \varepsilon \downarrow 0, \tag{7.28}$$

with parameters $D, \eta \in (0, \infty)$, makes it possible to derive an explicit LDP rate function for the normalized local times, which is in the spirit of the famous Donsker-Varadhan-Gärtner LDP [DonVar75, Gär77], see Lemma 4.1. The following is [KönSalWol12, Theorem 1.1]. We write $\mathbb{P}_0^\omega$ for the probability w.r.t. the random walk with conductances ω, starting from 0, and we write E and P for expectation and probability w.rt. the conductances.

Theorem 7.13 (Annealed LDP, finite region) *Let $B \subset \mathbb{Z}^d$ be a finite set containing the origin. Then, under the annealed measures $E[\mathbb{P}_0^\omega(\cdot\,|\mathrm{supp}\,(\ell_t) \subset B)]$, the normalised local times $\frac{1}{t}\ell_t$ satisfy a large deviation principle on the space $\mathcal{M}_1(B)$ of probability measures on $\mathbb{Z}^d$ with support in B with scale $t^{\frac{\eta}{\eta+1}}$ and rate function $J_0^{(\mathrm{d})} = J^{(\mathrm{d})} - \inf_{\mathcal{M}_1(B)} J^{(\mathrm{d})}$, where*

$$J^{(\mathrm{d})}(g^2) = K_{\eta,D} \sum_{e \in \mathcal{N}} \sum_{z \in \mathbb{Z}^d} \big|g(z+e) - g(z)\big|^{\frac{2\eta}{1+\eta}}, \qquad g^2 \in \mathcal{M}_1(B). \tag{7.29}$$

Here, $K_{\eta,D} = \big(1 + 1/\eta\big)(D\eta)^{1/(1+\eta)}$, and $\mathcal{N}$ is the set of positive neighbours of the origin.

Interesting is a comparison with the case of a non-random environment, corresponding to $\eta = \infty$. Indeed, in Theorem 7.13, we see an extrapolation of the rate function (the square of the ℓ^2-norm of the gradient) to the p-th power of the p-norm of the gradient for an arbitrary $p \in [1, 2)$; but note, however that the notation $\|\nabla g\|_p^p$ would be misleading, since it refers to the spatial function $\mathbb{Z}^d \ni y \mapsto g(y)$, but not to the coordinate function $\mathcal{N} \ni e \mapsto g(y+e)$ (here one takes the 2-norm by default).

In the case of a time-dependent growing box, interestingly there arise two cases, one of which, the case $\eta > d/2$, is just the continuous version of the fixed-box version, where one obtains a full LDP for the properly rescaled version of the local times. Here the local times spread out over the entire large box in a more or less homogeneous way. However, in the case $\eta < d/2$, it turns out in [KönWol15] that the LDP asymptotics follow the formulas of the fixed-box version, which seems to suggest that the random walk fills only a small part of the growing box.

Consider a spatial scaling function $\alpha_t \in (1, \infty)$ with $1 \ll \alpha_t \ll t^{1/2}$ and replace B by a time-dependent, growing set $B_t = \alpha_t G \cap \mathbb{Z}^d$, where we fix $G \subset \mathbb{R}^d$ as an open, connected and bounded set containing the origin and having a sufficiently regular boundary. In order to properly incorporate the t-dependence of the set B_t, like in Sect. 4.2, we consider the normalised and rescaled version L_t of ℓ_t, given by $L_t(x) = \frac{\alpha_t^d}{t}\ell_t(\lfloor \alpha_t x \rfloor)$. Recall that L_t is an L^1-normalised random step function on $\mathbb{R}^d$, having support in G on the event $\{\mathrm{supp}\,(\ell_t) \subset \alpha_t G\}$. A continuous analogue to the rate function in Theorem 7.13 is given by $J_0^{(c)} = J^{(c)} - \inf_{\mathcal{M}_1(G)} J^{(c)}$, where

$$J^{(c)}(f^2) = \begin{cases} K_{\eta,D} \sum_{i=1}^d \int_G \big|\partial_i f(y)\big|^{\frac{2\eta}{1+\eta}}\,\mathrm{d}y = K_{\eta,D} \sum_{i=1}^d \|\partial_i f\|_p^p, & \text{if } f \in H_0^1(G),\\ \infty, & \text{otherwise,}\end{cases} \tag{7.30}$$

where $p = \frac{2\eta}{1+\eta} \in [1,2)$, and $K_{\eta,D}$ is as in Theorem 7.13. The following is [KönWol15, Theorem 1.4]; we dropped some technical assumptions here.

Theorem 7.14 (Annealed LDP, time-dependent region) *Assume that $\eta > d/2$. Pick a scale function $(\alpha_t)_{t>0}$ such that $1 \ll \alpha_t^{d+2} \ll t(\log t)^{-(1+\eta)/\eta}$. Then the distributions of L_t under the conditional annealed measures $E[\mathbb{P}_0^\omega(\cdot \mid \mathrm{supp}\,(\ell_t) \subset \alpha_t G)]$ satisfies a large-deviation principle on $\mathcal{M}_1(G)$ with speed $\gamma_t = t^{\frac{\eta}{1+\eta}} \alpha_t^{\frac{d-2\eta}{1+\eta}}$ and good rate function $J_0^{(c)}$.*

It is also shown in [KönWol15] that the minimum of $J_0^{(c)}$ is attained for $\eta > d/2$, but not for $\eta \leq d/2$, in which case, in contrast, the discrete version, $J_0^{(c)}$, has a well-defined minimum. This shows that, for $\eta > d/2$, the random walk, given that it does not leave $\alpha_t G$ by time t, spreads out homogeneously over this area, and it seems to indicate (the proof is open yet) that for $\eta \leq d/2$, it collapses on a much smaller area inside $\alpha_t G$, possibly on some discrete subset that does not depend on t. Further studies will be necessary to understand this effect and to use the results of [KönSalWol12, KönWol15] for the study of the almost-sure setting and later for the study of the PAM with RWRC. However, here are some precise heuristics concerning the asymptotics of the annealed total mass of the solution, see [Wol13, Section 3.5]. They are based on a combination of the heuristics that we gave in Sect. 3.3, using Theorems 7.13 and 7.14.

Indeed, under Assumption (H) (see (3.26)) on the regularity of the upper tails of the potential single-site distribution (recall that $H(t)$ and $\eta(t)$ are regularly varying with parameter γ, see Proposition 3.15), we distinguish the cases that $\eta(t)$ is much larger than $t^{\eta/(1+\eta)}$ (implying that $\gamma \geq \eta/(1+\eta)$), or much smaller (implying that $\gamma \leq \eta/(1+\eta)$) or just on the scale $t^{\eta/(1+\eta)}$. Writing U^ω for the total mass with conductances ω and E for the expectation with respect to ω, we should then have in the first case, $\eta(t) \gg t^{\eta/(1+\eta)}$ just $\log E[\langle U^\omega(t)\rangle] \sim H(t)$, and in the second case, $\eta(t) \ll t^{\eta/(1+\eta)}$,

$$\log E[\langle U^\omega(t)\rangle] \sim \begin{cases} \eta(t)^a t^b \inf\limits_{f\in H^1(\mathbb{R}^d):\|f\|_2=1} \Big(\frac{\rho}{1-\gamma} \int |f|^{2\gamma} + J^{(c)}(f^2)\Big), & \text{if } \eta > d/2,\\ t^{\frac{\eta}{1+\eta}} \inf\limits_{g^2\in\mathcal{M}_1(\mathbb{Z}^d)} J^{(d)}(g^2), & \text{if } \eta \leq d/2,\end{cases} \tag{7.31}$$

where $a = a(\eta, \gamma) = 1/[d(1+\eta)(1-\gamma)]$ and $b = b(\eta, \gamma) = [d\eta(1-\gamma)]/[d\eta(1-\gamma) + 2\eta - d\gamma]$. However, in the last case, $\eta(t) \sim t^{\eta/(1+\eta)}$, one conjectures that

$$\log E[\langle U^{\omega}(t)\rangle] \sim -t^{\gamma} \inf_{g^2 \in \mathcal{M}_1(\mathbb{Z}^d)} \Big(\frac{\rho}{1-\gamma} \sum_{z \in \mathbb{Z}^d} |g(z)|^{2\gamma} + J^{(\mathrm{d})}(g^2) \Big), \tag{7.32}$$

i.e., the discrete-space version of the first line of (7.31). Note that there are now *three* random influences that might contribute on the leading scale in terms of a large-deviation behaviour. For instance, in case that the intermittent island has a t-dependent, unbounded size (i.e., if the underlying α_t diverges), a spatially continuous variational formula arises, and the influences of the potential and the conductances dominate that of the random walk.

It should be possible to prove the above conjectures along the line of some proofs of Theorem 3.13 available in the literature. The analysis of the variational formulas appearing seem to offer new interesting enterprises, and the investigation of the quenched settings is widely open and interesting. Note that there are three different quenched settings, depending on which randomness is kept fixed and averaged, respectively.

7.9.2 Localisation in the Bouchaud-Anderson Model

Recently [MuiPym14], the PAM was studied in another class of random environment, more precisely, the underlying random walk was replaced by the *Bouchaud random walk*, better known under the name *Bouchaud trap model*, where the randomness of the holding times does not sit in the bonds, but in the sites, and is chosen very heavy-tailed. More explicitly, the generator is given by

$$\Delta_{\sigma}^{\mathrm{d}} f(z) = \frac{1}{2d\sigma(z)} \sum_{y \in \mathbb{Z}^d : y \sim z} (f(y) - f(z)), \tag{7.33}$$

where $\sigma = (\sigma(z))_{z \in \mathbb{Z}^d}$, the *trapping landscape*, is a random i.i.d. field of positive numbers. The Bouchaud random walker, when standing in z, waits a random time that is exponentially distributed with expectation $2d\sigma(z)$ and then jumps with equal probability to any of the $2d$ neighbours. If $\log \sigma$ is Pareto-distributed (i.e., $\mathbb{P}(\sigma(0) > r) = (\log r)^{-\alpha}$ for all $r \geq 1$ for some $\alpha \in (0, \infty)$), then, almost surely with respect to the landscape σ, the Bouchaud random walk exhibits some peculiar behaviour that is caused by the existence of lattice sites with extra-ordinarily long holding time parameters; the random walk is trapped. This effect is most pronounced if the Pareto parameter α is smaller than one, and leads then to ageing phenomena. E.g., the trapping takes place after longer and longer time lags and in further and further remote sites, and both the time lags and the trapping sites show nice asymptotic scaling behaviours. See [BenČer06] for a summary of these properties of the Bouchaud random walk.

Combining the PAM with the Bouchaud random walk and considering concentration and ageing phenomena is tempting, as each of them shows these phenomena, the PAM at least for sufficiently heavy-tailed potential distributions, as we outlined in Sect. 6.4.2. Indeed, in [MuiPym14], the potential distribution is taken as Weibull-distributed, in which case these phenomena for the PAM are known from [FioMui14] and [SidTwa14]. However, the assumptions on the Bouchaud random walk made in [MuiPym14] are less restrictive, but do contain the Pareto case with parameter in $(0, 1)$.

The main result of [MuiPym14] is the complete localisation of the solution $u(t,\cdot)$ in one random, t-dependent site $Z_t \in \mathbb{Z}^d$ in the sense that $u(t,Z_t)/U(t)$ converges to one in probability. Furthermore, depending on the details of the distribution of potential, the potential values in the neighbouring sites of the concentration site Z_t are characterised; they indeed show an interesting limiting behaviour, and it can be seen in some sense that the one-site island slowly starts to emerge an interesting shape as the potential distribution gets less heavy-distributed.

7.10 Drifted PAM and Polymers

New interesting questions arise if a drift is added to the diffusion, i.e., if the generator $\Delta^{\rm d}$ of the simple random walk is replaced by the one of a random walk with drift. This changes the behaviour of the model and also its mathematical properties. One of the crucial issues is the loss of symmetry of the generator; a great deal of the mathematical tools presented in Sect. 2 does not apply. The Feynman-Kac formula does apply, but its large-deviations analysis is now quite different and has nothing to do with the Donsker-Varadhan-Gärtner LDP that we presented in Sect. 4.2.1. The underlying random walk has now a tendency to travel further and further and to see always new terrain, instead of returning to relatively few places many times and to build up high local times. It is not necessarily the highest potential values that play the leading rôle, but a compromise between quite high potential values and a multitude of regions that are successively relevant, as more and more time elapses.

Because of these substantial differences, we decided not to give an account on this subject, but to restrict to some fundamental remarks. The main conjecture about drifted PAM (or, equivalently, drifted random walk in random potential) is the existence of a *critical drift*, a critical threshold for the strength of the drift: for subcritical drifts, the behaviour of the total mass of the PAM is similar to the drift-free case, at least as it concerns the main terms in the asymptotics, and the intermittent islands should be essentially the same. In contrast, for supercritical drifts, the trajectory in the Feynman-Kac formula diverges at least with positive speed in the direction of the induced drift and develops a non-zero 'effective drift'.

This threshold between the two behaviours should be non-zero at least in the case of a bounded random potential, but not necessarily for unbounded ones, as extremely high peaks may suggest the path to spend much time in them, slowing down the speed.

Such critical drifts have been indeed established for a number of cases, more precisely, different behaviours have been found for sufficiently large and for sufficiently small drifts. In the former case for bounded potentials, large deviation principles for the end point of the random walk with linear scaling have been established, and criteria and bounds were given for the critical drift, both in the annealed and in the quenched setting.

However, most of the works on this subject offer no formula for the rate function nor for the critical drift, and all information derived about the behaviour of the motion and of the potential and about the critical drift is not quantitative; the validity of the LDP for the endpoint and the non-triviality of this transition is for many cases the main information that has been derived yet. The following questions are, to the best of my knowledge, widely open yet:

(i) Is there an additional intermediate regime between the two above mentioned ones (behaviour on the scale of the zero-drift case and non-trivial linear asymptotics)?
(ii) Can one characterise the critical drift(s) intrinsically and analytically?
(iii) On what scale does one see the influence of the induced drift in the zero-effective drift phase?

Closely connected with the drifted PAM is the question for the *Lyapounov exponent* of the motion in random potential. To define this, one replaces the deterministic time t in the Feynman-Kac formula (for *non-drifted* motion!) by the first time at which a certain hyperplane in a given direction $v \in \mathbb{R}^d$ with norm one with a certain distance r to the origin is reached. Then the Lyapounov exponent is defined as the large-r logarithmic asymptotics of this modified formula. It is interesting both in the annealed and in the quenched setting. Employing sub-additivity arguments of Kingman-type is not straight-forward, but require only technical work and can be considered standard by now. Results on the existence of Lyapounov exponents can easily be reformulated in terms of LDPs for the corresponding PAM with drifted motion. Deriving criteria for their non-triviality and for the validity of the above-mentioned transition needs some additional work.

For Brownian motion among Poisson obstacles, many deep results are presented in [Szn98], and there are versions for random walks in random potential under various assumptions on the distribution of the potential in the literature. We refer to [Flu07, Flu08] for some results in this direction and some guidance to the literature. One of the main questions is the coincidence of the annealed and the quenched Lyapounov exponents and their behaviours for vanishing diffusivity (i.e., for $\kappa \downarrow 0$, if the Hamiltonian $\kappa\Delta^{\mathrm{d}} + \xi$ is considered).

There seems to be a beginning of a story that derives and explores explicit formulas for drifted motions in random potential, see [Rue14], which might lead to a much deeper understanding and the investigation of various detailed questions in future.

The extreme case of drifts, the case where each step leads the trajectory by a fixed amount further in one direction, is called *directed polymer in random environment*, a name that reflects that this trajectory never hits a site more than once. (The term 'polymer' is often used in the mathematical literature for random paths that never, or at least rarely, produce self-intersections, like the *self-avoiding walk*.) Usually, one considers the time-discrete setting and picks the direction of the drift parallel to the first axis, such that the path that one considers is indeed of the form $(n, S_n)_{n\in\mathbb{N}_0}$ with $(S_n)_{n\in\mathbb{N}_0}$ a d-dimensional simple random walk (or other types of random walks). This is a $(1+d)$-dimensional polymer, which is indeed the graph of a d-dimensional walk.

Directed polymers in random environment are a subject of high importance, since they are believed to show behaviours that lie in the universality class of a number of prominent models, one of the most well-known of which is the *directed last-passage percolation* and the largest eigenvalue of a random matrix drawn from the *Gaussian unitary ensemble* and the *KPZ equation*; in all these models one observes fluctuations on the scale $t^{1/3}$ rather than $t^{1/2}$, a universal phenomenon of high importance that currently lacks a deeper understanding. We refer to the survey [ComShiYos04] from 2004 on directed polymers and the surveys [Cor12] and [Qua12] on the KPZ equation.

7.11 Branching Random Walks in Random Environment

In Sect. 2.1.1 we introduced the model of a branching random walk in a random environment of branching rates (BRWRE). The expected number of particles at time t in the site z is a solution to the PAM, where the expectation is taken over the branching/killing mechanism and the migration, but not over the branching rates, and the potential ξ is calculated from the branching and the killing rates. This fundamental relation suggests to exploit the knowledge on the PAM that has been gained over the last 20 years for the study of the BRWRE, but actually and a bit surprisingly, this has been done to a little extent yet. In this section, we report on these works.

The information on the branching process that one gains from the Feynman-Kac formula in (2.2) is indeed not very direct. The path $X = (X_s)_{s\in[0,t]}$ indeed stands for all the trajectories of branching particles that follow that path, and all branching and killing events are decoded in terms of the integral $\int_0^t \xi(X_s)\,ds$ over all ξ-values along the path in the exponent, but not in explicit branching/killing events. This makes it difficult to derive particlewise information about branching and killing. Nevertheless, the weight that the path X receives by the exponential integral should reflect the mass of particles that flow along that path.

From the understanding of the behaviour of the PAM, one can now guess that the behaviour of the BRWRE should have quite some features in common with the PAM. In particular, the arising picture should have nothing to do with a homogenised behaviour, and Brownian approximations in the sprit of Donsker's invariance principle should drastically fail. Instead, the branching particles should also enjoy an intermittency effect, i.e., they should be strongly concentrated in the same intermittent islands as the solution of the PAM is. Of high interest is then the description of the trajectories between these islands and the identification of the time scales and much more.

7.11.1 One-Dimensional Models

Some early work in that direction was carried out in a series of papers [GreHol92] for one-dimensional BRWRE in discrete time and space, with special attention to the influence of a drift to the expected number of particles, comparing and contrasting the annealed and quenched settings. No connection with the PAM (whose mathematical treatment was in its infancy at that time) was made, and the methods used there (Ray-Knight-type descriptions of the local time as a process in the space parameter) are strictly limited to one dimension. Hence, we are not going to present any details.

7.11.2 Moment Asymptotics for the Population Size

Motivated and influenced by [GärMol98], the d-dimensional continuous-time version of the BRWRE was studied in [AlbBogMolYar00] for Weibull-distributed branching rates with parameter $\alpha \in (1, \infty)$, such that the corresponding PAM lies in the class of double-exponentially distributed potentials with $\rho = \infty$, which is called the class (SP) of single-peak potentials in Sect. 3.4. The main focus was on deriving a Feynman-Kac-like formula for the expectation of the n-th power of the number of particles at time t at site z, $\eta(t, z)$, and the total particle number at time t, $\eta(t)$. This formula is presented there in a recursive fashion, which made it difficult to analyse its asymptotics as $t \to \infty$. The result identified the first term only, and it turned out that, for $p, n \in \mathbb{N}$,

$$\langle \mathrm{E}[\eta(t)^n]^p \rangle = \mathrm{e}^{H(npt)(1+o(1))}, \qquad t \to \infty.$$

That is, the asymptotics for the p-th moment (over the branching rates) of the n-th moment (over the branching/killing and migration, denoted by E) at time t are the same as the one of the first moment of the first moment at time tnp, at least as it concerns the first term. We know that this phenomenon can be easily interpreted for $n = 1$ (see Remark 3.2), but this is not so easy for arbitrary $n \in \mathbb{N}$.

This effect was later studied in greater detail and precision [GünKönSek13], where a direct version of that Feynman-Kac-like formula was derived, which admits deeper studies. The main tool there is the many-to-few lemma, an extension of the well-known many-to-one lemma from the theory of branching processes. For the branching rates doubly-exponentially distributed with any value of $\rho \in (0, \infty)$, also the second term in the asymptotics of $\langle \mathtt{E}[\eta(t)^n]^p \rangle$ was derived, and the above phenomenon is shown to hold true also for the second term (which is by the way, again given by the characteristic variational formula in (3.45)).

A closer inspection of the proof shows that this phenomenon should come from the issue that the potential ξ can attain positive values, and these values determine the asymptotics. However, for strictly negative potentials, the asymptotics of the p-th moment of the n-th moment should behave as the *first* moments at time tp, i.e., as if n would be one. The reason for this can be explained as follows. The above-mentioned Feynman-Kac-like formula for $\langle \mathtt{E}[\eta(t)^n]^p \rangle$ involves an expectation over a branching process that splits precisely $n-1$ times up to time t, and in the exponent there is a sum over all the integrals of the ξ-values over all the branches. The earlier all the $n-1$ branching times are, the longer the total time of the intervals that appear. If ξ can attain positive values, then this term is maximal if all the branching times are as early as possible, and if ξ attains only negative values, then they need to be as late as possible to produce a maximal value. Recalling that large-t asymptotics of exponential integrals over $[0, t]$ always come from a maximisation, this makes this effect plausible.

7.11.3 Intermittency for the Particle Flow

Motivated by the comprehensive understanding of the time-evolution of the PAM with Pareto-distributed potential, which we outlined in Sect. 6.5, [OrtRob14], [OrtRob16a] and [OrtRob16b] derived a similarly comprehensive picture of the time-evolution of the underlying continuous-time BRWRE on $\mathbb{Z}^d$ with Pareto-distributed branching rates. In particular, a precise and rigorous understanding of intermittency for the system is achieved.

Indeed, in [OrtRob14] it is shown that the branching particles are concentrated on the intermittent islands of the PAM (which are single sites now, see Sect. 6.4.2), but are traversed in a possibly different order than the main bulk of the mass of $u(t, \cdot)$ traverses it. This assertion is even strengthened in [OrtRob16b] by proving a one-point concentration property for the particles of the BRWRE. This proves an appealing ageing picture of the BRWRE in great detail. The main assertions are formulated in [OrtRob16a] in terms of a rescaled limit of the entire BRWRE model towards the *lilypad model*, a last-passage percolation process in continuous space with random, time-depending weights on the bonds between the points.

The main difference between the time-evolution of the main mass of the PAM and the main particle concentration of the BRWRE is the following. If the time t exceeds the threshold beyond which new, more preferable intermittent islands appear at the horizon, then the sample trajectory in the Feynman-Kac formula is completely rearranged from scratch and immediately walks from the origin to the new optimal potential site, without paying attention to the location of the last intermittent peak. The PAM searches for new islands only from the origin. In contrast, the main bulk of the branching particles is already located at the last intermittent island, and the particles have to travel from that location to a new one, which is now optimal as seen from that current location and does not have to be the one that the PAM-trajectory would choose.

Chapter 8
Time-Dependent Potentials

Of fundamental importance is the parabolic Anderson model in (1.1) also if the random potential is allowed to be time-dependent. Here we consider the Cauchy problem

$$\frac{\partial}{\partial t}u(t,z) = \kappa\Delta^{\mathrm{d}}u(t,z) + \xi(t,z)u(t,z), \qquad \text{for } (t,z) \in (0,\infty)\times\mathbb{Z}^d, \qquad (8.1)$$

$$u(0,z) = u_0(z), \qquad \text{for } z \in \mathbb{Z}^d, \qquad (8.2)$$

where $\xi\colon [0,\infty)\times\mathbb{Z}^d \to \mathbb{R}$ is a space-time random field that drives the equation, and u_0 is the initial datum, which we want to assume as a nonnegative and bounded function. This case is called the *dynamic* case, and the potential ξ is often called a *dynamic random environment* or a *dynamic potential*, in contrast with a *static* potential $\xi = (\xi(z))_{z\in\mathbb{Z}^d}$, which we considered in all the preceding chapters. Observe that the dynamic case is *not* the situation of a t-dependent scaling of a static potential, which we discussed in Sect. 7.3. The directed polymer in random environment, which we briefly mentioned in Sect. 7.10, belongs also to the dynamic setting, but the time dimension is interpreted as an additional space dimension there.

Usual general assumptions are that $\xi(0,0)$ is integrable and that the field ξ is time-space ergodic, which means in particular that the distribution of $\xi(\cdot,\cdot)$ is invariant under shifts both in time and space (and could therefore be extended also

W. König, *The Parabolic Anderson Model*, Pathways in Mathematics,
DOI 10.1007/978-3-319-33596-4_8

to negative times). The *Feynman-Kac formula* now reads

$$u(t,z) = \mathbb{E}_z\Big[\exp\Big\{\int_0^t \xi(t-s, X_s)\,\mathrm{d}s\Big\} u_0(X_t)\Big], \qquad z \in \mathbb{Z}^d, t \in (0,\infty), \tag{8.3}$$

where $(X_s)_{s\in[0,\infty)}$ is the continuous-time random walk on $\mathbb{Z}^d$ with generator $\kappa\Delta^{\mathrm{d}}$. Recall that the Feynman-Kac formula in the static case, (2.2), relies on a time-reversal, which made us consider paths moving from the initial site to z rather than from z to the initial site, but in (8.3) this would require a time-reversal of $\xi(t,\cdot)$. Actually, the finiteness of the right-hand side of (8.3) is already sufficient for the PAM in (8.1) to have a unique positive solution u. This is the extension of Theorem 1.2 to the dynamic case; see [CarMol94]. For the almost-sure existence of a solution, one has to find a criterion that ensures the finiteness of the right-hand side of (8.3) for almost all ξ; see e.g. Remark 8.11.

Like in the static case, which we considered in the preceding chapters, the main goal is the analysis of the solution $u(t,\cdot)$ of (8.1) in the large-t limit. For general time-dependent potentials, it is much more difficult to obtain qualitative information about the behaviour of the solution, and the results derived in the literature are much less explicit than in the static case. On the analytic side, the operator $\Delta^{\mathrm{d}} + \xi(t,\cdot)$ on the right-hand side of (8.1) depends on time, and therefore it is a priori not possible to make use of spectral theory here. In particular, an eigenvalue expansion is not possible here.

Also in (8.3), one sees that the picture is quite different from the static case. Again the large-t asymptotics should come from a behaviour of the path that spends as much time as possible in sites where the potential is extremely large. However, when the potential varies in time, this is much more difficult for the path, as the locations of the optimal regions may move. If the potential is mixing in time in any sense (or even i.i.d. in time, like for the directed polymers in random environment, which we briefly looked at in Sect. 7.10), then the fraction of time that the path spends in the extremal regions is very small, and the large-t behaviour of the solution is not likely to come from an extreme maximisation. Then the main contribution of the path to the Feynman-Kac formula will not come from many returns to a certain place, and hence the Donsker-Varadhan-Gärtner LDP theory will not be helpful.

Nevertheless, the concept of *intermittency* is as important as in the static case, but even more difficult to conceive (and to prove). In the spirit of the definition of moment intermittencyin (1.10), we call the solution *strongly p-intermittent* (sometimes just *p-intermittent*) for $p \in \mathbb{N}$ if either $\lambda_p(\kappa) = \infty$ or $\lambda_p(\kappa) > \lambda_{p-1}(\kappa)$, where the *annealed Lyapounov exponent* is defined as

$$\lambda_p(\kappa) = \lim_{t\to\infty} \frac{1}{t} \log\big[\langle u(t,0)^p\rangle^{1/p}\big], \tag{8.4}$$

provided these limits exist. We stress the dependence on the diffusion parameter κ; see the discussion in Remark 8.1. Like in the static case, a simple application of Jensen's inequality shows that $\lambda_p(\kappa)$ is non-decreasing in p, but proving the strict inequality needs some efforts. Unlike in the static case, it is generally unknown whether the strict inequality $\lambda_p(\kappa) > \lambda_{p-1}(\kappa)$ for some p implies it for every p, hence the notion a priori needs to depend on p. This notion may appear meaningless if all the $\lambda_p(\kappa)$ are equal to ∞, which may happen for unbounded potentials, see Example 8.3. Certainly, this phenomenon is expected to have something to do with a concentration property of the solution in small, far apart regions, but such a property is rather difficult to coin explicitly, even more difficult to prove, and possibly not true, even if the Lyapounov exponents were strictly increasing in p; this is largely unexplored. Negative results have been derived for such a concentration property for potentials that are mixing in time, see Remark 8.11. In this respect, the research still is close to the beginning, and much insight in the geometry of the solution $u(t,\cdot)$ is still lacking.

Also the rôle of the *quenched Lyapounov exponent*,

$$\lambda_0(\kappa) = \lim_{t\to\infty} \frac{1}{t} \log u(t,0), \tag{8.5}$$

for the geometric properties of the main mass flow is not yet understood, but there are some results for its existence and dependence on the diffusion constant κ, see the discussion in Remark 8.1 and some results in Remark 8.11 and in Sect. 8.3.

We are now giving a survey on results on the large-time behaviour of the solution for a number of interesting random potentials. We concentrate on the annealed Lyapounov exponents in Sect. 8.1 and on the quenched one in Sect. 8.2; While in Sect. 8.1 we consider mainly examples from population dynamics, we briefly turn to white-noise potentials in Sect. 8.3, but keep it short, as we encounter the boundaries of this book's scope there, just when we are about to enter another exciting research area.

8.1 Potentials Built Out of Particles

One type of random potentials $\xi(t,\cdot)$ that received much interest in recent years as the driving potential for the PAM is built out of a given family of random processes in $\mathbb{Z}^d$. More precisely, $\xi(t,z)$ is (up to a constant) equal to the number of these random walkers at time t at the site z. This has a convincing interpretation as a family of *catalysts* for the *reactants*, whose expected number is registered in the solution to the PAM in (8.1). We remind on the discussion in Remark 2.1, the only difference being that the catalyst particles are here evolving in time. If $\xi_*(t,z)$ denotes their number at time t in the site z, and if each reactant splits into two with rate $\gamma \in (0,\infty)$ per catalyst and dies with rate $\delta \in [0,\infty)$, then their expected number at time t at site z is equal to $u(t, z)$ (the solution to (8.1)) with $\xi = \gamma\xi_* - \delta$. See also Sects. 2.1.1 and 7.11 for the static case.

In a series of papers by Gärtner, den Hollander and Maillard, the following types of catalyst particles are considered:

(0) One single random walk with generator $\rho\Delta^{\mathrm{d}}$, starting from the origin,
(i) Independent random walks (ISRW) with generator $\rho\Delta^{\mathrm{d}}$, starting from a Poisson random field (that is, in every lattice point the number of catalysts at the beginning is independently Poisson distributed, say with parameter $\nu > 0$).
(ii) Symmetric exclusion process (SEP): At each time, every site is either occupied by one particle or empty. Particles jump from a site x to a neighbouring site y at rate $p(x, y) = p(y, x) \in (0, 1)$, if y is empty.
(iii) Symmetric voter model (SVM): At each time, every site is either occupied by one particle or empty. Site x imposes its state on a neighbouring particle at y at rate $p(x, y) = p(y, x) \in (0, 1)$.

While in model (0), the catalyst is localised and ξ therefore not shift-invariant in distribution, in models (i)–(iii), the catalysts form a large family that covers large parts of the space. The initial distribution of catalyst particles is such that the corresponding process is in equilibrium, i.e., the numbers of particles in the sites forms a stationary process in time. In models (i) and (ii), it is even reversible, which makes an analysis via spectral theory and large-deviations analysis possible, but the arising variational formulas are rather involved and difficult to analyse further.

We summarize results on the model (0) in Example 8.2. The annealed results found by that team for the models (i)–(iii) in a series of papers are summarized in Examples 8.3, 8.5, and 8.6, see also the survey article [GärHolMai09b].

Remark 8.1 (What we can learn from the annealed Lyapounov exponents) The annealed and the quenched Lyapounov exponents in (8.4) and (8.5) are the quantities that give the first non-trivial information about the large-t behaviour of the PAM. The information that one can draw from them is, on the one hand, much more descriptive than the first asymptotic term in the static case (which just gives the asymptotics of $\langle \mathrm{e}^{t\xi(0)}\rangle$), but, on the other hand, much less descriptive and detailed than the second term in the static case. In general, the Lyapounov exponents cannot be identified in terms of interpretable variational formulas that admit a detailed further analysis. However, they do give some information about the question whether or not the reactant (if necessary, jointly with the catalysts) follows a particular strategy to optimise its contribution to the Feynman-Kac formula or not. Here we mean a strategy that costs probabilistically on the exponential scale and is made up for by a larger gain of the interaction between them, i.e., the term in the exponent of the Feynman-Kac formula.

For a catalyst-potential like the above four models, a large value of $\lambda_p(\kappa)$ comes from a clumping behaviour of the catalysts, in combination with a close nestling of the reactant to the catalysts. (In contrast, if γ would be taken negative, then a good joint strategy would be that the catalysts and the reactant try to avoid each other as much as possible; see Remark 8.8.) If both the catalyst and the reactant would not follow such a cooperative joint strategy and would behave just 'as usual', then the Lyapounov exponent would be just equal to the value that one would obtain if the potential ξ is replaced by its expected value everywhere. In the case where the expected value does not depend on time nor on space (e.g., if ξ is assumed stationary in time and space), then a simple application of Jensen's inequality shows that $\lambda_p(\kappa) \geq \langle \xi(0,0) \rangle$. The interesting question is whether this inequality is strict, since it would say that the cooperative strategy of catalysts and the reactant is noticeable on the exponential scale. This question could be settled in great generality, see below.

One of the main interests lies in the behaviour of the annealed Lyapounov exponent $\lambda_p(\kappa)$ as a function of κ, again in a comparison to the expected value of ξ. One interesting question is whether or not $\lambda_p(\kappa)$ is decreasing in κ and what its limiting value as $\kappa \to \infty$ is: the expected value or larger? The monotonicity in κ, at least for the four above types of potentials, should come from the fact that they are built out of random walks on $\mathbb{Z}^d$, like the reactant in the Feynman-Kac formula. This is the more difficult (i.e., probabilistically costly), the larger the diffusion constant κ of the reactant is. The interesting question then is whether or not anything of this joint strategy stays over in $\lambda_p(\infty)$ or not. ◇

Let us report on some results for the model (0).

Example 8.2 (Finitely many catalysts) Let $Y = (Y_t)_{t\in[0,\infty)}$ be a random walk on $\mathbb{Z}^d$ with generator $\rho\Delta^{\mathrm{d}}$ starting from the origin, and put $\xi = \gamma\xi_* - \delta$, where $\xi_*(t,z) = \mathbb{1}_{\{Y_t=z\}}$, i.e., the potential is equal to one in the current location of the catalyst and zero anywhere else. The initial condition is localised, i.e., $u(0,\cdot) = \delta_0(\cdot)$. Hence, catalyst and reactant both start from the origin. The growth of the expected mass of the reactants will be produced precisely in the site where the catalyst is. The asymptotics now depend heavily of the question how costly it is that the two particles (catalyst and reactant) travel most of the time together in the same site. This has been addressed in [GärHey06] and [SchWol12].

Key to analysing the exponential growth of the first moment is a spectral analysis of the operator $\kappa\Delta^{\mathrm{d}} + r\delta_0$. This suggests itself, as the analytic description of the above strategy in terms of the local time profile of the reactant (more precisely, of the process X in the Feynman-Kac formula) amounts to a delta-like shape, and the peak can be thought of as being at the origin. It comes therefore as no surprise that in [GärHey06] the Lyapounov exponent $\lambda_p(\kappa)$ turns out to be equal to ρ times the upper boundary of the spectrum of $\kappa\Delta^{\mathrm{d}} + r\delta_0$ with the choice $r = p\gamma/\rho$. Therefore, in analytical terms, exponential first moment growth corresponds to a positive spectral radius of that operator, which is in turn equivalent with the fact that $r\delta_0$ is a compact perturbation of $\kappa\Delta^{\mathrm{d}}$, which is itself equivalent to the existence of a positive eigenfunction. This is known to be true in $d = 1$ and $d = 2$ for any value of $r > 0$, but in $d \geq 3$ only for r sufficiently large. Correspondingly, we observe in $d \geq 3$ an interesting dependence of the question of intermittency on the parameter γ/ρ, while in $d \leq 2$ intermittency occurs for any p. The interpretation is that in higher dimensions, it is more difficult to follow the catalyst than in low dimensions, and he needs a stronger incentive for doing that.

Reference [SchWol12] makes much more precise assertions about the behaviour of the solution $u(t,\cdot)$ close to the catalyst. Indeed, they describe the solution from the view point of the particle by looking at $u(t, Y_t + \cdot)$ and derive asymptotics for its annealed moments even up to equivalence. For higher moments, they first derived the spectral analysis of a suitable modification of the operator $\kappa\Delta^{\mathrm{d}} + r\delta_0$.

In the case of multiple catalysts (but finitely many) and/or higher moments, the perturbing term that appears in the analytical description of the annealed Lyapunov exponent is not compact any more, which adds an additional degree of complexity to the problem. Annealed asymptotics for multiple catalysts are treated in [CasGünMai12]. ◇

In the following examples, we report on annealed results on the choices (i)–(iii) of models.

Example 8.3 (Independent simple symmetric random walks) Let us describe some annealed results for model (i) from [GärHol06]. They put the killing rate $\delta = 0$ and look at $\xi = \gamma\xi_*$ with $\xi_*(t,z)$ equal to the number of catalysts at time t in the site z. Here, the random potential ξ is unbounded to ∞, as many of the simple random walks can be present at a given site, even for long time lags. In particular, the probabilistic costs for the ISRW to have of order e^t, say, particles close to the origin for most of the time interval $[0,t]$ might be made up for by the gain of the reactant from staying in that area for long time and enjoying a large production rate. Then the Feynman-Kac formula would be then even of order $\mathrm{e}^{\mathrm{e}^t}$. This scenario is similar to the static case with potentials from the class (DE) or (SP), but on a much larger scale.

It is shown in [GärHol06] that the limit in (8.4) exists for any $\kappa \in [0,\infty)$ and that the map $\kappa \mapsto \lambda_p(\kappa)$ is finite, continuous, non-increasing and convex on $[0,\infty)$ if $\lambda_p(0)$ is finite. Furthermore, introducing the Green's function $G_d = \int_0^\infty p_t(0,0)\,\mathrm{d}t$ (with $p_t(x,y)$ the transition probability of one of the ISRWs), $\lambda_p(\kappa)$ is finite for all κ if and only if $p < 1/G_d\gamma$. Since $G_1 = G_2 = \infty$, the solution u is, in dimensions $d = 1$ and $d = 2$, therefore p-intermittent for any p and, in $d \geq 3$, for all sufficiently large p. This means in particular that the expected number of reactants per site goes to ∞ super-exponentially fast in $d = 1$ and $d = 2$. Such a behaviour is called *strongly catalytic* in [GärHol06]. See Remark 8.4 for similar results derived independently relying on path estimates.

However, for $d \geq 3$, the Lyapounov exponent is finite only for sufficiently small p, depending on γ. For $p > 1/G_d\gamma$, it shown that the divergence of $\frac{1}{t}\log[\langle u(t,0)^p\rangle^{1/p}]$ is even exponential in t, i.e., the expectation of the solution runs on an double-exponential scale, as we indicated above. For $p < 1/G_d\gamma$, [GärHol06] identifies the values of $\lambda_p(0)$ and the asymptotics of $\kappa\lambda_p(\kappa)$ for $\kappa \to \infty$, and they show that $\lambda_p(\kappa)$ is now even strictly decreasing in κ. It turns out that $\lambda_p(0)$ is strictly larger than γ times the mean number of ISRWs everywhere (the expectation of ξ) and that $\lambda_p(\kappa)$ decreases to that value as $\kappa \to \infty$.

However, the question whether or not the model is p-intermittent (i.e., whether $\lambda_p(\kappa) > \lambda_{p-1}(\kappa)$) is left open there; they conjecture that this is true for any κ in $d = 3$, but only for sufficiently small κ in $d \geq 4$. Interestingly, in $d = 3$, the asymptotics of $\kappa\lambda_p(\kappa)$ show a remarkable connection with the polaron model, whose mean-field version we briefly mentioned in Example 7.11; there is a heuristics for deeper reasons behind it. This three-dimensional effect makes a difference in the second-order term of the convergence of $\lambda_p(\kappa)$, which gives rise to the conjecture. ◇

Remark 8.4 (Survival and extinction of branching random walks with catalysts) The interpretation of interacting reactants and catalysts in the model (i) above has also been studied in [KesSid03] with the additional assumption that reactant particles die at a certain deterministic rate $\delta \in (0, \infty)$. Independently of [GärHol06], the authors make the intriguing observation that, in dimensions 1 and 2, the expected number of reactants at a site grows to infinity in time regardless of the choice of the other model parameters, including the killing rate δ. Additionally, the corresponding growth rate is always faster than exponential. These are annealed results, which also follow from the results of [GärHol06] that we summarize above.

Furthermore, conditioning on the behaviour of the catalysts (i.e., looking at the quenched situation), the reactants die out in all dimensions if δ is chosen large enough. This interesting behaviour indicates that there are immensely high peaks in the concentration of reactants along the space of different reealisations of the catalyst behaviour. However, the reactants locally survive if δ is small enough, which shows an interesting phase transition. ◇

Example 8.5 (Symmetric exclusion process) Let us describe the results of [GärHolMai07] and [GärHolMai09a] for the model (ii). The symmetric exclusion process is a family of independent, identically distributed random walks in $\mathbb{Z}^d$ whose transition kernel is symmetric and irreducible, subject to the exclusion rule: no site can ever be occupied by more than one particle, i.e., all jump decisions towards an occupied site are suppressed. Recall that $\xi = \gamma\xi_*$, where $\xi_*(t,z) \in \{0,1\}$ is the number of catalysts at the site z at time t.

It turns out that the annealed Lyapounov exponent $\lambda_p(\kappa)$ defined in (8.4) exists, is finite, non-increasing and convex in κ, for all values of the parameters. There is an interesting dychotomy in the recurrent behaviour of the catalysts. Indeed, if the underlying random walk that drives the exclusion process is recurrent, then $\lambda_p(\kappa)$ is equal to the value that one would have if ξ_* would be replaced by one everywhere (i.e., its maximal value), while in the transient case, it is strictly below, but still strictly larger than the expected value of the potential ξ. However, as $\kappa \to \infty$, it approaches the latter value, and similar asymptotics for the difference are derived as for the case described in Example 8.3, including the interesting connection with the polaron term and the conjectures about intermittency. The latter question could be answered in the affirmative only for $\kappa = 0$ for any p.

The interpretation is that, in the recurrent case, the main asymptotics comes from a clumping of the catalysts in a large ball for a long time, i.e., the catalysts fill practically every site in that ball, while in the recurrent case, such a ball is only filled gradually with a higher rate than the mean value. This effect vanishes as the diffusion strength of the reactant is driven to ∞. ◇

Example 8.6 (Symmetric voter model) Here ξ_* is given as a system of 'opinions' 0 or 1 at each site in $\mathbb{Z}^d$, and the person at x imposes his/her opinion to the person at y with rate $p(x, y) \in [0, 1]$ (if they are different), where p is the transition kernel of an irreducible symmetric random walk on $\mathbb{Z}^d$. Two different initial states are considered, the Bernoulli process or the equilibrium measure, each with density (expected opinion in a site) $\rho \in (0, 1)$. Again, we take $\xi = \gamma\xi_*$ as the random potential.

In [GärHolMai10], the existence and finiteness of the Lyapounov exponent $\lambda_p(\kappa)$ in (8.4) is proved and its continuity in κ. For the following, assume that the steps under the kernel $p(0, \cdot)$ have a finite variance. Then the following dychotomy is identified: for $d \leq 4$, the value of $\lambda_p(\kappa)$ is the same as if all opinions would be equal to one, while for $d > 4$, it is strictly between that value and the expected value of the potential, i.e., the one that one would get if all opinions would be replaced by their mean value, ρ. Furthermore, in $d > 4$, the solution is shown to be p-intermittent for any p for $\kappa = 0$. It is only a conjecture that, in these dimensions, the Lyapounov exponent approaches the expected value of the potential, and that it does this in a very particular manner that again involves an interesting variational problem in $d = 5$, like the polaron problem in $d = 3$. Hence, the knowledge is less developed than in Examples 8.3 and 8.5 above, which is partially due to the fact that the ISRW and the SEP are reversible and admit some spectral-theoretic methods, but the SVM is not. ◇

Remark 8.7 (Random conductances) In [ErhHolMai15b], the simple random walk is replaced by the random walk among random conductances (RWRC), i.e., the operator $\kappa\Delta^{\mathrm{d}}$ is replaced by the randomised Laplace operator $\Delta^{\mathrm{d}}{}_{\omega}f(z) = \sum_{x\sim z}\omega_{x,z}(f(x) - f(z))$, see Sect. 7.9.1 for the static case. That is, the *random conductance model* is considered. The random conductances are assumed to be uniformly elliptic (bounded and bounded away from zero) and symmetric, i.e., attached to the undirected edges. They do not have to be i.i.d. over the edges, but only need to satisfy a certain clustering property. The random time-dependent potential ξ is assumed to be either an i.i.d. field of Gaussian white noises (see Sect. 8.3), or a field of (finitely or infinitely many) independent random walks, or a spin-flip system. Then the Lyapounov exponent, almost surely with respect to the conductances, is shown to exist and to coincide with the supremum of the Lyapounov exponents with conductances $\equiv \kappa$, taken over all κ in the support of the conductances. ◇

Let us have a brief look at the situation with negative potential:

Remark 8.8 (Randomly moving traps) We obtain a model for reactant movement among randomly moving *traps* if we choose $\gamma < 0$. In this case, the solution $u(t, z)$ under the initial condition $u(0, \cdot) = \delta_0$ describes the *survival probability* of a randomly moving particle that is killed at rate $-\gamma$ times the number of traps present at the same site.

The case of a single randomly moving trap $Y = (Y(t))_{t\in[0,\infty)}$, i.e., $\xi(t, z) = \gamma\delta_{Y_t}(z)$ with $\gamma \in [-\infty, 0]$, is not accessible by the spectral theory of the operator

$\kappa\Delta^{\mathrm{d}} + r\delta_0$ with negative r (see Example 8.2), as the spectrum of the Laplacian is concentrated on the negative half-axis and a perturbation by $r\delta_0$ does not create an isolated positive eigenvalue. Precise large time asymptotics for the expected total mass of the solution have been treated in [SchWol12] for the localised initial condition $u(0,\cdot) = \delta_0(\cdot)$. Indeed, the asymptotics are not on any exponential scale, but on the scale of the Green's function of the random walk. That is, the expectation of $\sum_{z\in\mathbb{Z}^d} u(t,z)$ runs like a constant times $t^{-1/2}$ in $d = 1$, a constant times $1/\log t$ in $d = 2$, and it even converges in $d \geq 3$. Also the initial condition $u(0,\cdot) \equiv 1$ is considered in [SchWol12]. Here the expected total mass converges in any dimension, but only in $d = 1$ towards some non-trivial term, otherwise just to one.

In [DreGärRamSun12], the ISRW as in model (i) at the beginning of this section (see Example 8.3) are considered, however, with negative γ. More precisely, instead of the Feynman-Kac formula in (8.3), [DreGärRamSun12] considers

$$\widetilde{U}(t) = \mathbb{E}_0\Big[\exp\Big\{\gamma\int_0^t \xi_*(s,X_s)\,\mathrm{d}s\Big\}\Big], \qquad z\in\mathbb{Z}^d, t\in(0,\infty), \tag{8.6}$$

with ξ_* as in Example 8.3, and $\gamma \in [-\infty, 0)$. The main result of [DreGärRamSun12] is the following annealed asymptotics:

$$\log\langle\widetilde{U}(t)\rangle \sim \begin{cases} -\nu\sqrt{\frac{8\rho t}{\pi}}, & \text{if } d = 1,\\ -\nu\pi\rho\frac{t}{\log t}, & \text{if } d = 2,\\ -\lambda_1(\kappa)t, & \text{if } d \geq 3, \end{cases}$$

where $\lambda_1(\kappa)$ is a number that is not characterised further, but it is shown that it satisfies $\lambda_1(\kappa) \geq \lambda_1(0)$. Further work on the behaviour of the path X in (8.6) is done in [AthDreSun16] for $d = 1$; in particular it is shown that it is subdiffusive, i.e., X_t runs like $o(\sqrt{t})$. See also [DreGärRamSun12] for some more biographical information about models of trapping paths by other particles. ◇

8.2 The Quenched Lyapounov Exponent

We turn now to the quenched Lyapounov exponent $\lambda_0(\kappa)$ defined in (8.5). Results on this have been derived for general space-time ergodic random potentials $\xi\colon [0,\infty)\times\mathbb{Z}^d$ under various additional assumptions, and they have been specialised to the four examples of Sect. 8.1. Our report on these results will not be exhaustive, but the work [GärHolMai12] (see Remark 8.9 below) makes contributions to quenched results and summarizes much of the literature about the small-κ behaviour of the quenched Lyapounov exponent $\lambda_0(\kappa)$.

Also the quenched Lyapounov exponent gives rise to some interesting effects. First, if the potential ξ is unbounded from above, the finiteness of $\lambda_0(\kappa)$ is not at all clear, as in principle it is possible that the reactant can make substantial use of the high peaks of ξ, spend much time in them without paying too much probabilistically and producing a large, possibly super-exponentially large, value of the Feynman-Kac formula. However, if the potential is sufficiently mixing in time and space, it can be shown that this is not the case, see Remark 8.11 below.

The question about the qualitative behaviour of the map $\kappa \mapsto \lambda_0(\kappa)$ is also highly interesting, as it says something about how well the reactant can make use of the high peaks of the potential in dependence of the strength of the diffusivity, in particular, when just 'switching on' its movement, i.e., for small κ. It is expected that this map starts from the expected value of ξ at $\kappa = 0$, increases for a while and then decays back to the expected value, but this has not been settled yet in a satisfactory way.

Remark 8.9 (Quenched results for ISRW, SEP and SVM) In [GärHolMai12], the authors analyse the quenched asymptotics (i.e., conditioned on the evolution of catalyst particles), more precisely, the quenched Lyapounov exponent $\lambda_0(\kappa)$ defined in (8.5). They do this for the localised solution (i.e., $u(0,\cdot) = \delta_0(\cdot)$) and first under the assumption of stationarity and ergodicity of ξ with respect to translations in $\mathbb{Z}^d$ and that, for all value of the parameters, $\langle \log u(t,0)\rangle \leq ct$ for all t, where the constant c may depend on the parameters. In a second step, they specialise the results to the three examples ISRW, SEP and SVM that we outlined in Examples 8.3, 8.5 and 8.6. The bound $\langle \log u(t,0)\rangle \leq ct$ is clearly satisfied for bounded potentials like SEP and SVM, but also for ISRW, as [KesSid03] shows.

The start is that $\lambda_0(\kappa)$ exists almost surely and in L^1-sense and is finite. Then they show that it is globally Lipschitz-continuous outside any neighbourhood of 0 and that it is strictly larger than the expected value of ξ, $\gamma\rho$. Under some weak technical condition (which is satisfied for the three examples), $\lambda_0(\kappa)$ is shown not to be Lipschitz-continuous at 0. For some other types of potentials, this is substantiated by some explicit upper bound for $\lambda_0(\kappa) - \gamma\rho$ in the limit $\kappa \downarrow 0$, which is of order $\log\log \frac{1}{\kappa} / \log \frac{1}{\kappa}$. For the three examples, it is shown that $\lambda_0(\kappa)$ tends to $\gamma\rho$ as $\kappa \to \infty$, and for SEP and SVM, there is a lower bound for $\lambda_0(\kappa) - \gamma\rho$ in the limit $\kappa \downarrow 0$ of order $1/\log \frac{1}{\kappa}$. ◇

Remark 8.10 (More general quenched results) In [DreGärRamSun12], the existence and finiteness of a certain modification of the quenched Lyapounov exponent is established (both in almost-sure sense and in L^1) for space-time ergodic bounded potentials ξ that are reversible in time or symmetric in space, the modification being that the time-reversal of ξ in the Feynman-Kac formula (8.3) is dropped and (8.6) is considered instead. The novelty in comparison to [GärHolMai12] (see Remark 8.9) is that this limit is shown not to depend on the (non-negative and bounded) initial condition u_0, even if u_0 has an unbounded support.

In [ErhHolMai14], a very general potential $\xi\colon [0,\infty) \times \mathbb{Z}^d \to \mathbb{R}$ (taking possibly positive and negative values and being possibly unbounded) is considered, and some basic properties of the quenched Laypounov exponent are derived, which extend those of [GärHolMai12]. Indeed, ξ is assumed to be space-time ergodic such that

$\xi(0,0)$ is integrable, and that, for any $T > 0$, the field $\xi^{(T)}(\cdot) = \sup_{t\in[0,T]} \xi(t,\cdot)$ is percolating from below. The latter means by definition that each of its level sets $\{z \in \mathbb{Z}^d : \xi^{(T)}(z) \leq \alpha\}$ with $\alpha > 0$ contains an infinite connected set. Under these assumptions, it is proved in [ErhHolMai14] that the Feynman-Kac formula in (8.3) is finite for any bounded nonnegative initial condition u_0 and therefore the solution to (8.1) is unique. Furthermore, if a certain space-time mixing property of ξ is assumed (called *Gärtner-mixing* in [ErhHolMai14]), then the existence and finiteness of the quenched Lyapounov exponent $\lambda_0(\kappa)$ is established, its independence on the initial datum u_0, and its continuity in $\kappa \in [0,\infty)$ and Lipschitz continuity away from the origin. Recall from Remark 8.1 that the finiteness of $\lambda_0(\kappa)$ is a non-trivial issue if ξ is unbounded. ◇

Here is an interesting *negative* result on 'quenched' intermittency in a general setting. We remarked earlier that p-intermittency (defined in terms of the moments) should have quite something to do with the relevance of high peaks in the random potential for the solution, but is rather difficult to nail down. Even worse, for some interesting cases in which this notion of intermittency is satisfied, the almost-sure behaviour of the solution does not reflect the characteristics of a relevance of high peaks, or at least only in a very mild form, as we want to report on now.

Remark 8.11 (Moment-intermittency versus almost-sure intermittency) The results of [ErhHolMai14] (see Remark 8.10), interesting as they are on their own, are only the preparation for a much deeper result that is derived in the follow-up paper [ErhHolMai15a]. Here a stronger mixing property (called *Gärtner-hyper-mixing*) is imposed, under which it is shown that (at least for the localised initial condition $u_0 = \delta_0$) $\lim_{\kappa\to\infty} \lambda_0(\kappa) = \langle \xi(0,0)\rangle$. This is even more interesting as there are examples of Gärtner-hyper-mixing potentials for which the annealed Lyapounov exponent $\lambda_1(\kappa)$ is infinite for any $\kappa \in [0,\infty)$, i.e., 1-intermittency holds. Hence, even though this indicates that high peaks in the potential landscape are relevant for the large-t behaviour of the moments, their influence on the almost-sure behaviour even vanishes if the diffusivity is sent to ∞. ◇

8.3 White Noise Potential

Another very natural choice of a time-dependent random potential is a *Gaussian white-noise potential* on $\mathbb{Z}^d$, which can be formally written as $\xi(t,z) = \frac{\partial}{\partial t} W_t^{(z)}$, where $(W^{(z)})_{z\in\mathbb{Z}^d}$ is a collection of independent standard Brownian motions. Another formal definition is that $(\xi(t,z))_{z\in\mathbb{Z}^d, t\in[0,\infty)}$ is a centred Gaussian field with covariance function $\langle \xi(t,z)\xi(s,x)\rangle = \delta_0(s-t)\delta_0(z-x)$. This potential is not a function, but a distribution, but nevertheless it is possible to give a good sense to the solution [CarMol94].

Reference [CarMol94] commenced a thorough study of the PAM with this potential, and the high degree of independence and the Gaussian nature made it possible to establish the existence and finiteness of all the Lyapounov exponents, both the annealed and the quenched ones. Furthermore, it was possible to initiate a study of the dependence of these exponents on κ, where the limit as $\kappa \downarrow 0$ is most interesting. For technical reasons, [CarMol94] only work for an initial datum u_0 that is nonnegative and has a compact support. Indeed, in [CarMol94] and [GreHol07], the authors establish the following picture: p-intermittency is present in the recurrent cases $d = 1, 2$ for any $\kappa \geq 0$ and any $p \in \mathbb{N}$, whereas in $d \geq 3$, it occurs only if κ is small enough, depending on p and d.

More precisely, all annealed Lyapounov exponents exist and are finite, and they are continuous and non-increasing in κ and converge towards the expected potential value 0 as $\kappa \to \infty$. In $d = 1$ and $d = 2$, all the curves $\kappa \mapsto \lambda_p(\kappa)$ are strictly decreasing in κ and do never reach zero and they are always strictly ordered in p. Hence, p-intermittency holds for any p. In $d \geq 3$, for any p, the function $\kappa \mapsto \lambda_p(\kappa)$ starts from some strictly positive value, decays strictly until it reaches 0 at some value $\kappa_p > 0$ and then stays at zero. The critical values κ_p are strictly increasing in p. Furthermore, for $\kappa \in [0, \kappa_p)$, we have $\lambda_p(\kappa) > \lambda_{p-1}(\kappa)$ and therefore p-intermittency. Let us remark that, in [BorCor12], an explicit formula for the moments for the solution to the PAM at fixed time was derived, even for any *drifted* nearest-neighbour Laplacian in terms of contour integrals in the complex plane, which results into an explicit formula for all the p-th Lyapounov exponents.

The asymptotic behaviour of the quenched Lyapunov exponent $\lambda_0(\kappa)$ as $\kappa \to 0$ has been analysed in [CarMol94] for the case where the initial datum u_0 has compact support. We refer to [GärHolMai12] for further references and literature remarks about questions like the extension to more general initial data, the asymptotics of $\lambda_0(\kappa)$ as $\kappa \downarrow 0$, and whether or not $\kappa_1 < \kappa_2$ or about the behaviour of $\lambda_p(\kappa)$ for $p \downarrow 0$.

The spatially continuous analogue is also highly natural and technically very demanding and poses a number of further questions for the future, not only with respect to large-t asymptotics and intermittency. Here, the random potential $\xi = (\xi(t,x))_{t\in[0,\infty),x\in\mathbb{R}^d}$ is a Gaussian space-time white noise, i.e., a centred Gaussian field with covariance $\langle \xi(t,x)\xi(s,y)\rangle = \delta_0(t-s)\delta_0(x-y)$. Certainly, this is only a formal expression, and ξ is indeed a distribution with low regularity. Similarly to the brief argument in Example 1.21, one sees that the construction of a solution has a chance only in dimension one. Even more, in all the other dimensions, one has no hope to construct a solution even with the use of the recently established methods that we mentioned there. Hence, we have to restrict to $d = 1$.

In this case, the PAM is often called the *stochastic heat equation* and is written

$$\partial_t u = \Delta u + \xi u \qquad \text{on } [0,\infty) \times \mathbb{R},$$

with a space-time white noise ξ. Formally, there is an interesting representation as the *Cole-Hopf-transformation* $u(t,x) = \mathrm{e}^{h(t,x)}$ of the solution to the *KPZ equation*

$$\partial_t h = (\partial_x h)^2 + \partial_x^2 h + \xi,$$

which is of great interest because it appears in many models in a universal way and is believed to exhibit a highly interesting asymptotic behaviour as it concerns the order of the fluctuations. Actually, the KPZ equation and related models are currently one of the hottest research subjects in probability and related fields, but clearly outside the scope of this text. This topic recently received a particular push when the theory of regularity structures was developed so far that a solution to the KPZ equation could be constructed [Hai13]; see also [HaiLab15b].

The diversity of the literature about the PAM and the KPZ equation has started to explode, as a lot of interesting questions are addressed, like mollified modifications of the white noise, discrete-space approximations, Lévy white noise, large-x properties of the solution for fixed t, discrete approximations with rescaled equations, and much more, see e.g., [Che15a, Che15b, CraGauMou10, CraMou06]. We refer the reader to [Cor12] and to [Qua12] for detailed surveys on the KPZ equation and adjacent topics.

Appendix: Open Problems

In this appendix, we give a little guidance to some interesting open research projects and research directions in the vicinity of the PAM. These items have been mentioned already in this text, but they are scattered, and we felt it would be useful to collect them at one place. The list is certainly influenced by our personal taste. We do not repeat here areas that are too far away from the main body of this text, like random walk in random scenery (Sect. 7.4), self-attractive path measures (Sect. 7.5), drift and directed polymers (Sect. 7.10), and time-dependent potentials (Chap. 8). Certainly, if the PAM is put into a much broader connection, a lot of new exciting research areas open up, like the consideration of more general types of PDEs with random coefficients, or the addition of terms that introduce new physical effects, or other classes of time-depending potentials. However, recall that the scope of this book is defined by the characteristics that

- the solution admits an explicit solution,
- the analysis of its long-time behaviour can be based on large-deviation analysis and extreme-value statistics,
- the arising variational formulas are explicit and admit interpretations and deeper investigation,
- the solution shows a clear geometric picture,
- there is a clear connection with the spectrum of the Anderson operator.

For many interesting PDEs with random coefficients, already the solution theory is challenging, and it cannot be hoped for deriving a clear picture in near future; the research on them is entirely different from the study of the PAM as we reported on in this text. Let us give a list of open problems that we find interesting.

Other potential distributions. The PAM has not yet been analysed for some interesting potential distributions that are popular in the study of the spectrum of the Anderson Hamiltonian, like alloy-type potentials (see Example 1.19). Furthermore, random potentials with long correlations (see Sect. 7.2), in particular the Gaussian free field both on $\mathbb{Z}^d$ and on $\mathbb{R}^d$ and, more generally, log-correlated random fields,

W. König, *The Parabolic Anderson Model*, Pathways in Mathematics,
DOI 10.1007/978-3-319-33596-4

are currently much studied with respect to their local and global maxima. They show some interesting new structures like certain relations to branching processes and Poisson process descriptions with random intensity measures. Hence, it appears rather interesting to initiate a study of the PAM with such fields as the random input and to investigate how the correlation lengths may result into a new intermittency picture featuring new classes of intermittent islands (larger asymptotics of their radii, new variational formulas) or even a transition to a homogenised behaviour, possibly after inserting some extra changes, like some time-dependent pre-factors. See the last remark in Sect. 7.3 for a general ansatz for the study of the PAM with correlated fields, which seems to be particularly suitable for many Gaussian fields on $\mathbb{Z}^d$.

The PAM with a Gaussian white-noise potential in the spatially continuous setting, see Example 1.21, is currently very much in fashion. One reason is that the PAM presents one of the few important and prominent types of stochastic partial differential equations to which the florishing theory of regularity structures, rough paths and other related methods on the edge between analysis and probability theory is applied. The interest stems from the facts that (1) the continuous-space PAM with Gaussian white noise should arise as the rescaling limit of the discrete-space PAM in the spirit of Donsker's invariance principle, (2) the construction of a solution is not possible in all the dimensions and requires a subtle smoothing and rescaling, and (3) for future investigations with respect to long-time features like ageing and intermittency, the construction has to fulfill high requirements. Currently, the latest achievements are constructions of solutions to the PAM in selected dimensions in the full space, but only locally in time, and the mathematical foundation of the spectral theory of the Anderson Hamiltonian is in its infancy. The big enterprise will be to see how far the new techniques can be extended. Furthermore, the study of the local maxima of the Gaussian white noise will certainly lead to other results than the study of maxima of the local eigenvalues of the Anderson Hamiltonian (unlike in the case of a smooth Gaussian potential).

Eigenvalue order statistics, one-island concentration and time evolution. The most comprehensive and detailed description of the mass flow modeled by the PAM can be given in terms of a concentration property of the solution in just one of the intermittent islands (see Sect. 6.4) and an description of its location as a function of time (see Sect. 6.5). This in turn requires a full control on the behaviour of all the top eigenvalues and eigenfunctions of the Anderson Hamiltonian as given in terms of a Poisson process approximation; see Sect. 6.3. This programme has been carried out for practically all heavier-tailed potentials (in the sense of Example 1.14) and for the double-exponential distribution of Example 1.12. However, it is still open for such important distributions like the Bernoulli traps or, more generally, bounded potentials, also in the spatially continuous setting, i.e., Poisson traps, neither for any kind of Gaussian field. At the end of Sect. 6.3.3 we briefly pointed out that we expect that much of the techniques derived in [BisKön16] and [BisKönSan16] will be useful. But a substantial new input will be necessary to overcome problems related to the characteristic feature that small changes in the eigenvalue will not come from larger values of the potential, but larger sizes of the intermittent island. For general Gaussian potentials in the continuous setting, working out each of these points seems to be widely open and promising, even in the classical case where a high degree of regularity is assumed, like in [GärKönMol00]. For deriving such detailed information as an order statistics, a very precise and explicit control on the distribution of the potential seems necessary.

Anderson localisation via local maximisation. A further, quite ambitious, question is about the geometric interpretation of Anderson localisation deeper in the spectrum of the Anderson operator $\Delta^{\mathrm{d}} + \xi$. With the help of spatial versions of extreme-value statistics, one was able to characterise local areas of concentration of the leading eigenfunctions in large boxes and to derive a kind of Anderson localisation picture from that, as we explained in Sect. 6.3.4. However, one knows from much less explicit methods developed in the community of Anderson localisation that much more eigenfunctions are localised, not only the leading ones. We think it should be highly interesting to derive a geometric characterisation also of high local peaks in areas away from the highest potential peaks, which give rise to localisation of eigenfunctions away from the leading ones. The goal would be to use methods from extreme-value theory to describe such structures of local potential maxima and their influence on the spectrum of $\Delta^{\mathrm{d}} + \xi$. Since these methods would depend on an analysis in large boxes, one cannot strictly speak of Anderson localisation, hence, the next step must be to relate the findings in the large-box setting to the spectral properties in $\mathbb{Z}^d$.

Transition between concentration and homogenisation. As we reported on in Sect. 7.3, interesting critical regimes and phase transitions and variational formulas arose in the study of the PAM with an accelerated motion or, equivalently, a weakly interacting potential. A deep study was carried out for Brownian motion in a scaled Poisson trap field only (see Sect. 7.3.3), but in the general case on $\mathbb{Z}^d$, a variational formula (see (7.15)) was derived that might contain a phase transition in great generality if the parameter θ ranges from small to large values: the formula should have no solutions for small θ and should be compact for large ones. This has not yet been explored, and it should be done both on the level of the variational analysis and the behaviour of the PAM. Different behaviours in the respective dimensions should arise, and it will be interesting to find criteria on H (i.e., on the random potential) for them to hold. Intimately connected is the study of the (top of the) spectrum of $\Delta^{\mathrm{d}} + \varepsilon\xi$ in large, ε-depending boxes of various choices of the radii $\gg \varepsilon^{-2}$ in the limit $\varepsilon \downarrow 0$.

PAM in random environment. As we discussed in Sect. 7.9.1, another interesting enterprise is the PAM in random environment, i.e., when the simple random walk is replaced by a random walk in random environment. One natural choice is the random walk among random conductances (RWRC). Here, already some precise heuristics have been formulated (see Sect. 7.9.1), but yet only for the behaviour of the moments. Deeper insight and possibly the introduction of new methods will be necessary if one wants to study any of the almost sure settings, where one takes the average over the potential only, the environment only, or none of them. The RWRC is the easiest and most comfortable random environment to study, since it admits still the exploitation of a well-functioning ℓ^2-theory and an explicit variational analysis, because there is a symmetric generator. The study becomes much more challenging if instead a general random walk in random environment is considered. As a (already highly intriguing) pre-study, its long-time behaviour in boxes, possibly with slowly diverging radius, is interesting, i.e., an extension of the work reported on in Sect. 7.9.1 to non-symmetric random walks.

PAM with stable diffusivity. Replacing the driving motion (simple random walk and Brownian motion) by some random motion in the domain of attraction of stable processes, or replacing the Laplace operator by some fractional version of it, gives rise to new, interesting formulas and effects, both on the probabilistic and the analytic side. As we reported in Sect. 7.8, we are aware only of one work in that direction, and it is clear that the non-continuity of the paths, respectively the non-locality of the operator, give rise to additional effects that wait for exploration.

PAM with drift. In spite of a substantial interest in drifted random motions in random potential, most of the relevant works rely on subadditivity and do therefore not offer any explicit formulas, see Sect. 7.10. A notable exception is [Rue14] for the spatially continuous setting, i.e., Brownian motion in a quite general random potential. Here the logarithmic long-distance and the large-time asymptotics are expressed in terms of a variational formula involving the well-known energy term $\|\nabla\varphi\|_2^2$ and another one that describes the influence of the potential. The interpretation of these terms for the motion is not direct and requires some manipulations, but nevertheless it is there and presumably will give rise to some interesting discoveries, as soon as one undertakes efforts to make this relation more explicit. Also the understanding of the quenched setting will greatly benefit from a deeper analysis of the relation between the variational formula and suitable objects encoded in the Feynman-Kac formula.

More realistic biological population models. One of the main interpretations and applications of the PAM is in terms of spatial stochastic particle systems with branching and killing, as we remarked at some places, notably in Remark 2.1.1. However, in those cases of a static random potential that can produce large branching rates, the growth of the population is ridiculously large in the long-time limit, and the contrast between single sites with such a gigantic offspring production and the ample regions around with almost no growth is far beyond all reasonably observed real population histories. One obvious drawback of this model for the explanation of population histories is the absence of any kind of birth control, even at places where the local population is enormous. There is a lack of reasonable population models in random environments that include such effects, but can still be handled mathematically. One possibility is to combine spatial versions in random environment of the Moran model, which is well-known in biological stochastic modeling, where the number of individuals is kept fixed over the entire duration of the process, or of the Lenski experiment, where the population is randomly thinned out after certain time lags. However, it seems as if no substitute for the Feynman-Kac formula is in sight for such models.

Asymptotics for the (non-parabolic, time-dependent) Anderson Schrödinger equation. The Anderson Schrödinger equation in (1.4), i.e., the non-parabolic version, has been shown in [Wag14] to be amenable to a probabilistic analysis, see Remark 2.2. Indeed, the papers by Wagner seem to have opened the door to a comprehensive analysis of the Schrödinger equation with the help of the theory of spatial marked branching processes. This makes possible the application of a powerful probabilistic toolbox. Looking at the representation of the solution in (2.1), the biggest mathematical obstacle seems to be to find a useful Feynman-Kac formula (possibly on an enlarged state space, e.g., $\mathbb{Z}^d \times \{-1, 1\} \times \{+, -\})$ and to master the technical problems coming from the difference of the particle numbers, causing a lot of extinction.

Bibliography

[Adl90] R.J. ADLER, *An Introduction to Continuity, Extrema, and Related Topics for General Gaussian Processes.* Hayward: Inst. Math. Stat. (1990). 95

[AthDreSun16] S. ATHREYA, A. DREWITZ and R. SUN, Subdiffusivity of a random walk among a Poisson system of moving traps on $\mathbb{Z}$. preprint (2016). 167

[AlbBogMolYar00] S. ALBEVERIO, L.V. BOGACHEV, S.A. MOLCHANOV and E.B. YAROVAYA, Annealed moment Lyapunov exponents for a branching random walk in a homogeneous random branching environment. *Markov Proc. Relat. Fields* **6**, 473–516 (2000). 155

[AllCho15] R. ALLEZ and K. CHOUK, The continuous Anderson hamiltonian in dimension two, preprint (2015). 18

[And58] P.W. ANDERSON, Absence of diffusion in certain random lattices. *Phys. Rev.* **109**, 1492–1505 (1958). 33

[Ant94] P. ANTAL, *Trapping Problems for the Simple Random Walk.* Dissertation ETH Zürich, No. 10759 (1994). 61, 80, 91

[Ant95] P. ANTAL, Enlargement of obstacles for the simple random walk. *Ann. Probab.* **23:3**, 1061–1101 (1995). 61, 80

[AssCas03] A. ASSELAH and F. CASTELL, Large deviations for Brownian motion in a random scenery. *Probab. Theory Relat. Fields* **126**, 497–527 (2003). 79

[AssCas07] A. ASSELAH and F. CASTELL, Self-intersection times for random walk, and random walk in random scenery in dimensions $d \geq 5$. *Probab. Theory Relat. Fields* **138:1–2**, 1–32 (2007). 141

[Ast08] A. ASTRAUSKAS, Extremal theory for spectrum of random discrete Schrödinger operator. I. Asymptotic expansion formulas. *J. Stat. Phys.* **131:5**, 867–916 (2008). 111

[Ast12] A. ASTRAUSKAS, Extremal theory for spectrum of random discrete Schrödinger operator. II. Distributions with heavy tails. *J. Stat. Phys.* **146:1**, 98–117 (2012). 111

[Ast13] A. ASTRAUSKAS, Extremal theory for spectrum of random discrete Schrödinger operator. III. Localization properties. *J. Stat. Phys.* **150:5**, 889–907 (2013). 111

[Ast16] A. ASTRAUSKAS, From extreme values of i.i.d. random fields to extreme eigenvalues of finite-volume Anderson Hamiltonian. preprint (2016). 101, 111

[Ban80] C. BANDLE, *Isoperimetric inequalities and applications.* Monographs and Studies in Mathematics, vol. 7, Pitman, Boston, Mass. (1980). 64, 95

W. König, *The Parabolic Anderson Model*, Pathways in Mathematics,
DOI 10.1007/978-3-319-33596-4

[BasCheRos06] R. BASS, X. CHEN and J. ROSEN, Moderate deviations and laws of the iterated logarithm for the renormalized self-intersection local times of planar random walks. *Electron. J. Probab.* **11**, 993–1030 (2006). 137

[BecKön12] M. BECKER and W. KÖNIG, Self-intersection local times of random walks: exponential moments in subcritical dimensions. *Probab. Theory Relat. Fields* **154:3–4**, 585–605 (2012). 81

[BenBogMol05] G. BEN AROUS, L. BOGACHEV and S. MOLCHANOV, Limit theorems for sums of random exponentials. *Probab. Theory Relat. Fields* **132**, 579–612 (2005). 146

[BenČer06] G. BEN AROUS and J. ČERNÝ, Dynamics of trap models. Math. Stat. Physics Lecture Notes – Les Houches Summer School **83** (2006). 152

[BenKipLan95] O. BENOIS, C. KIPNIS and C. LANDIM, Large deviations from the hydrodynamic limit of mean zero asymmetric zero range processes. *Stoch. Proc. Appl.* **55:1**, 65–89 (1995). 75

[BenMolRam05] G. BEN AROUS, S. MOLCHANOV and A. RAMIREZ, Transition from the annealed to the quenched asymptotics for a random walk on random obstacles. *Ann. Probab.* **33**, 2149–2187 (2005). 146

[BenMolRam07] G. BEN AROUS, S. MOLCHANOV and A. RAMIREZ, Transition asymptotics for reaction-diffusion in random media. In: *Probability and Mathematical Physics: A Volume in Honor of Stanislav Molchanov*, AMS/CRM, 42, 1–40 (2007). 146

[BerBolHol01] M. VAN DEN BERG, E. BOLTHAUSEN and F. DEN HOLLANDER, Moderate deviations for the volume of the Wiener sausage. *Ann. of Math. (2)* **153:2**, 355–406 (2001). 139, 143

[BerBolHol04] M. VAN DEN BERG, E. BOLTHAUSEN and F. DEN HOLLANDER, On the volume of the intersection of two Wiener sausages. *Ann. of Math. (2)* **159**, 741–782 (2004). 143

[BerBolHol05] M. VAN DEN BERG, E. BOLTHAUSEN and F. DEN HOLLANDER, Brownian survival among Poissonian traps with random shapes at critical intensity. *Probab. Theory Related Fields* **132:2**, 163–202 (2005). 139, 143

[BinGolTeu87] N.H. BINGHAM, C.M. GOLDIE and J.L. TEUGELS, *Regular Variation*. Cambridge University Press (1987). 60

[Bis11] M. BISKUP, Recent progress on the Random Conductance Model. *Prob. Surveys* **8**, 294–373 (2011). 148

[BisFukKön16] M. BISKUP, R. FUKUSHIMA and W. KÖNIG, Eigenvalue fluctuations for lattice Anderson Hamiltonians. preprint (2016). 135

[BisKön01] M. BISKUP and W. KÖNIG, Long-time tails in the parabolic Anderson model with bounded potential. *Ann. Probab.* **29:2**, 636–682 (2001). 55, 61, 64, 65, 77, 88, 89, 91, 92, 93

[BisKön01a] M. BISKUP and W. KÖNIG, Screening effect due to heavy lower tails in one-dimensional parabolic Anderson model. *Jour. Stat. Phys.* **102:5/6**, 1253–1270 (2001). 92

[BisKön16] M. BISKUP and W. KÖNIG, Eigenvalue order statistics for random Schrödinger operators with doubly exponential tails. *Commun. Math. Phys.* **341:1**, 179–218 (2016). 60, 101, 107, 109, 110, 111, 174

[BisKönSan16] M. BISKUP, W. KÖNIG and R. DOS SANTOS, Mass concentration in one island for the parabolic Anderson model with doubly exponential tails. in preparation (2016). 60, 101, 114, 119, 174

[Bol86] E. BOLTHAUSEN, Laplace approximations for sums of independent random vectors. *Probab. Theory Relat. Fields* **72:2**, 305–318 (1986). 125

[Bol94] E. BOLTHAUSEN, Localization of a two-dimensional random walk with an attractive path interaction. *Ann. Probab.* **22**, 875–918 (1994). 125

[Bol02] E. BOLTHAUSEN Large deviations and interacting random walks. Lectures on probability theory and statistics (Saint-Flour, 1999), 1-124, Lecture Notes in Math., 1781, Springer, Berlin (2002). 137, 144

[BolHol94] E. BOLTHAUSEN and F. DEN HOLLANDER, Survival asymptotics for Brownian motion in a Poisson field of decaying traps. *Ann. Probab.* **22:1**, 160–176 (1994). 143

[BolKönMuk15] E. BOLTHAUSEN, W. KÖNIG and C. MUKHERJEE, Mean-field interaction of Brownian occupation measures, II: Rigorous construction of the Pekar process, preprint (2015). 84, 144

[BolSch97] E. BOLTHAUSEN and U. SCHMOCK On self-attracting d-dimensional random walks, *Ann. Probab.* **25:2**, 531–572 (1997). 59, 137, 144

[BorCor12] A. BORODIN and I. CORWIN, Moments and Lyapunov exponents for the parabolic Anderson model. *Ann. Appl. Prob.* **24:3**, 1172–1198 (2014). 170

[Bra02] A. BRAIDES, *Gamma-convergence for Beginners*, Oxford University Press (2001). 48

[BryHofKön07] D. BRYDGES, R. VAN DER HOFSTAD and W. KÖNIG, Joint density for the local times of continuous time Markov chains. *Ann. Probab.* **35:4**, 1307–1332 (2007). 80, 81

[BrySla95] D. BRYDGES and c G. Slade, The diffusive phase of a model of self-interacting walks. *Probab. Theory Relat. Fields* **103:3**, 285–315 (1995). 137

[CarLac90] R. CARMONA and J. LACROIX, *Spectral Theory of Random Schrödinger Operators*, Probability and its Applications, Birkhäuser Boston (1990). 34, 35, 36

[CarMol94] R. CARMONA and S.A. MOLCHANOV, Parabolic Anderson problem and intermittency. *Mem. Amer. Math. Soc.* **108** no. 518 (1994). 2, 5, 6, 19, 160, 169, 170

[CarMol95] R. CARMONA and S.A. MOLCHANOV, Stationary parabolic Anderson model and intermittency. *Probab. Theory Relat. Fields* **102**, 433–453 (1995). 69, 96

[Cas10] F. CASTELL, Large deviations for intersection local time in critical dimension, *Ann. Probab.* **38:2**, 927–953 (2010). 81

[CasGünMai12] F. CASTELL, O. GÜN, and G. MAILLARD, Parabolic Anderson Model with a finite number of moving catalysts. In: *Probability in Complex Physical Systems*, Festschrift for Erwin Bolthausen and Jürgen Gärtner, Springer Proceedings in Mathematics Volume 11, 2012, DOI: 10.1007/978-3-642-23811-6, Heidelberg (2012). 164

[CasLauMél14] F. CASTELL, C. LAURENT and MÉLOT, Exponential moments of self-intersection local times of stable random walks in subcritical dimensions, *J. Lond. Math. Soc. (2)* **89:3**, 876–902 (2014). 81, 82

[Che10] X. CHEN, *Random Walk Intersections: Large Deviations and Related Topics.* Mathematical Surveys and Monographs, AMS (2010) Vol. 157, Providence, RI. 72, 137, 142

[Che12] X. CHEN, Quenched asymptotics for Brownian motion of renormalized Poisson potential and for the related parabolic Anderson models. *Ann. Probab.* **40**, 1436–1482 (2012). 66

[Che14] X. CHEN, Quenched asymptotics for Brownian motion in generalized Gaussian potential, *Ann. Probab.* **42**, 576–622 (2014). 16, 69, 96

[Che15a] X. CHEN, Spatial asymptotics for the parabolic Anderson models with generalized time-space Gaussian noise, to appear in *Ann. Probab.*, preprint (2015). 171

[Che15b] X. CHEN Precise intermittency for the parabolic Anderson equation with an $(1+1)$-dimensional time-space white noise, to appear in *Annales de l'Institut Henri Poincaré*, preprint (2015). 171

[CheKul12] X. Chen and A.M. Kulik, Brownian motion and parabolic Anderson model in a renormalized Poisson potential, *Annales de l'Institut Henri Poincare* **48**, 631–660 (2012). 15, 16, 24, 66, 139

[CheKul11] X. Chen and A.M. Kulik, Asymptotics of negative exponential moments for annealed Brownian motion in a renormalized Poisson potential, *International Journal of Stochastic Calculus, Art. Int. J. Stoch. Anal.* **43**, Art. ID 803683, 2090–3340 (2011). 66, 140

[CheXio15] X. Chen and J. Xiong, Annealed asymptotics for Brownian motion of renormalized potential in mobile random medium. *J. Theor. Probab.* **28**, 1601–1650 (2015). 24, 140

[CheRos11] X. Chen and J. Rosinski, Spatial Brownian motion in renormalized Poisson potential: A critical case. preprint (2011) 24

[ComShiYos04] F. Comets, T. Shiga, and N. Yoshida, Probabilistic analysis of directed polymers in a random environment: a review. *Advanced Studies in Pure Mathematics* **39**, 115–142 (2004). 154

[Cor12] I. Corwin, The Kardar-Parisi-Zhang equation and universality class. *Random Matrices: Theory and Applications* **1** (2012). 154, 171

[CraGauMou10] M. Cranston, D. Gauthier and T.S. Mountford, On large deviations for the parabolic Anderson model. *Probab. Theory Relat. Fields* **147**, 349–378 (2010). 171

[CraMou06] M. Cranston and T.S. Mountford, Lyapunov exponent for the parabolic Anderson model in $\mathbb{R}^d$. *J. Funct. Anal.* **236:1**, 78–119 (2006). 171

[CraMol07] M. Cranston and S.A. Molchanov, Quenched to annealed transition in the parabolic Anderson problem. *Probab. Theory Relat. Fields* **138:1–2**, 177–193 (2007). 146

[DemZei98] A. Dembo and O. Zeitouni, *Large Deviations Techniques and Applications.* 2nd Edition. Springer, New York (1998). 47, 56, 57, 67, 72

[DonVar75] M. Donsker and S.R.S. Varadhan, Asymptotics for the Wiener sausage. *Comm. Pure Appl. Math.* **28**, 525–565 (1975). 30, 66, 75, 76, 132, 149

[DonVar75-83] M. Donsker and S.R.S. Varadhan, Asymptotic evaluation of certain Markov process expectations for large time, I–IV. *Comm. Pure Appl. Math.* **28**, 1–47, 279–301 (1975), **29**, 389–461 (1979), **36**, 183–212 (1983). 73, 75

[DonVar79] M. Donsker and S.R.S. Varadhan, On the number of distinct sites visited by a random walk. *Comm. Pure Appl. Math.* **32**, 721–747 (1979). 75

[DonVar83] M. Donsker and S.R.S. Varadhan, Asymptotics for the polaron. *Comm. Pure Appl. Math.* **36**, 505–528 (1983). 144

[DreGärRamSun12] A. Drewitz, J. Gärtner, A. Ramirez, and R. Sun, Survival probability of a random walk among a Poisson system of moving traps. In: *Probability in Complex Physical Systems*, Festschrift for Erwin Bolthausen and Jürgen Gärtner, Springer Proceedings in Mathematics Volume 11, 2012, DOI: 10.1007/978-3-642-23811-6, Heidelberg (2012). 167, 168

[Dyn88] E.B. Dynkin, Self-intersection gauge for random walks and for Brownian motion. *Ann. Probab.* **16**, 1–57 (1988). 81

[Eis95] N. Eisenbaum, Une version sans conditionnement du théorème d'isomorphisme de Dynkin. In *Séminaire de Probabilités, XXIX.* Lecture Notes in Math. **1613**, 266–289. Springer, Berlin (1995). 81

[ErhHolMai14] D. Erhard, F. den Hollander and G. Maillard, The parabolic Anderson model in dynamic random environment: basic properties of the quenched Lyapunov exponent. *Ann. Inst. Henri Poicaré* **50:4**, 1231–1275 (2014). 168, 169

[ErhHolMai15a] D. Erhard, F. den Hollander and G. Maillard, The parabolic Anderson model in dynamic random environment: space-time ergodicity for the quenched Lyapunov exponent. *Probab. Theory Relat. Fields* **162**, 1–46 (2015). 169

[ErhHolMai15b] D. Erhard, F. den Hollander and G. Maillard, Parabolic Anderson model in a dynamic random environment: random conductances, preprint (2015). 148, 166

[FioMui14] A. Fiodorov and S. Muirhead, Complete localisation and exponential shape of the parabolic Anderson model with Weibull potential field. *Elec. J. Probab.* **19(58)**, 1–27 (2014). 101, 111, 118, 121, 152

[FleMol90] K. Fleischmann and S. Molchanov, Exact asymptotics in a mean-field model with random potential. *Probab. Theory Relat. Fields* **86**, 239–251 (1990). 8

[Flu07] M. Flury, Large deviations and phase transition for random walks in random non-negative potentials. *Stoch. Proc. Appl.* **117**, 596–612 (2007). 153

[Flu08] M. Flury, Coincidence of Lyapunov exponents for random walks in weak random potentials. *Ann. Probab.* **36:4**, 1528–1583 (2008). 153

[Fuk08] R. Fukushima, Asymptotics for the Wiener sausage among Poissonian obstacles. *Jour. Stat. Phys.* **133:4**, 639–657 (2008). 126

[Fuk09a] R. Fukushima, Brownian survival and Lifshitz tail in perturbed lattice disorder. *Jour. Func. Ana.* **256:9**, 2867–2893 (2009). 17

[Fuk09b] R. Fukushima, From the Lifshitz tail to the quenched survival asymptotics in the trapping problem. *Elec. Comm. Prob.* **14**, paper 42, 435–446 (2009). 35

[Fuk11] R. Fukushima, Second-order asymptotics for Brownian motion in a heavy-tailed Poissonian potential. *Markov Proc. Relat. Fields* **17:3**, 447–482 (2011). 66, 133

[FukUek10] R. Fukushima and N. Ueki, Classical and quantum behavior of the integrated density of states for a randomly perturbed lattice. *Ann. Inst. Henri Poincaré, Probab. Stat.* **11:6**, 1053–1083 (2010). 17

[FukUek11] R. Fukushima and N. Ueki, Moment asymptotics for the parabolic Anderson problem with a perturbed lattice potential. *Jour. Func. Anal.* **260:3**, 724–744 (2011). 17

[GanKönShi07] N. Gantert, W. König and Z. Shi, Annealed deviations for random walk in random scenery. *Ann. Inst. Henri Poincaré (B) Prob. Stat.* **43:1**, 47–76 (2007). 74, 79

[Gär77] J. Gärtner, On large deviations from the invariant measure. *Th. Prob. Appl.* **22**, 24–39 (1977). 73, 75, 149

[GärHey06] J. Gärtner and M. Heydenreich, Annealed asymptotics for the parabolic Anderson model with a moving catalyst. *Stoch. Process. Appl.* **116:11**, 1511–1529 (2006). 163

[GärHol99] J. Gärtner and F. den Hollander, Correlation structure of intermittency in the parabolic Anderson model. *Probab. Theory Relat. Fields* **114**, 1–54 (1999). 59, 60, 62, 102, 125, 126, 128, 129, 144

[GärHol06] J. Gärtner and F. den Hollander, Intermittency in a catalytic random medium. *Ann. Probab.* **34:6**, 2219–2287 (2006). 164, 165

[GärHolMai07] J. Gärtner, F. den Hollander and G. Maillard, Intermittency on catalysts: symmetric exclusion. *Elec. J. Prob.* **12**, 516–573 (2007). 165

[GärHolMai09a] J. Gärtner, F. den Hollander and G. Maillard, Intermittency on catalysts: three-dimensional simple symmetric exclusion. *Elec. J. Prob.* **14**, 2091–2129 (2009). 165

[GärHolMai09b] J. Gärtner, F. den Hollander and G. Maillard, Intermittency on catalysts. in: J. Blath, P. Mörters and M. Scheutzow (eds.), *Trends in Stochastic Analysis*, London Mathematical Society Lecture Note Series 353, Cambridge University Press, Cambridge, pp. 235–248 (2009). 162

[GärHolMai10] J. Gärtner, F. den Hollander and G. Maillard, Intermittency on catalysts: voter model. *Ann. Probab.* **38:5**, 2066–2102 (2010). 166

[GärHolMai12] J. GÄRTNER, F. DEN HOLLANDER and G. MAILLARD, Quenched Lyapunov exponent for the parabolic Anderson model in a dynamic random environment. In: *Probability in Complex Physical Systems*, Festschrift for Erwin Bolthausen and Jürgen Gärtner, Springer Proceedings in Mathematics Volume 11, 2012, DOI: 10.1007/978-3-642-23811-6, Heidelberg (2012). 167, 168, 170

[GärKön00] J. GÄRTNER and W. KÖNIG, Moment asymptotics for the continuous parabolic Anderson model. *Ann. Appl. Probab.* **10:3**, 192–217 (2000). 16, 55, 60, 69, 77, 95, 131

[GärKön05] J. GÄRTNER and W. KÖNIG, The parabolic Anderson model. in: J.-D. Deuschel and A. Greven (Eds.), *Interacting Stochastic Systems*, pp. 153–179, Springer (2005). 6

[GärKönMol00] J. GÄRTNER, W. KÖNIG and S. MOLCHANOV, Almost sure asymptotics for the continuous parabolic Anderson model. *Probab. Theory Relat. Fields* **118:4**, 547–573 (2000). 16, 60, 95, 97, 132, 175

[GärKönMol07] J. GÄRTNER, W. KÖNIG and S. MOLCHANOV, Geometric characterization of intermittency in the parabolic Anderson model. *Ann. Probab.* **35:2**, 439–499 (2007). 52, 60, 62, 93, 101, 102, 103, 104, 105, 109, 116

[GärMol90] J. GÄRTNER and S. MOLCHANOV, Parabolic problems for the Anderson model I. Intermittency and related topics. *Commun. Math. Phys.* **132**, 613–655 (1990). 2, 5, 6, 10, 11, 19, 20, 22

[GärMol98] J. GÄRTNER and S. MOLCHANOV, Parabolic problems for the Anderson model II. Second-order asymptotics and structure of high peaks. *Probab. Theory Relat. Fields* **111**, 17–55 (1998). 13, 14, 45, 55, 60, 61, 62, 74, 88, 89, 92, 108, 109, 155

[GärMol00] J. GÄRTNER and S. MOLCHANOV, Moment asymptotics and Lifshitz tails for the parabolic Anderson model. Stochastic models (Ottawa, ON, 1998), 141–157, CMS Conf. Proc., 26, Amer. Math. Soc., Providence, RI (2000). 130, 131

[GärSch11] J. GÄRTNER and A. SCHNITZLER, Time correlations for the parabolic Anderson model. *Elec. Jour. Probab.* **16**, 1519–1548 (2011). 122

[GärSch15] J. GÄRTNER and A. SCHNITZLER, Stable limit laws for the parabolic Anderson model between quenched and annealed behaviour. *Ann. Inst. Henri Poincaré, Probab. Stat.* **51:1**, 194–206 (2015). 146

[GerKlo11] F. GERMINET and F. KLOPP, Spectral statistics for the discrete Anderson model in the localized regime. *RIMS Kōkyūroku Bessatsu* **B27**, 11–24 (2011). 112

[GerKlo13] F. GERMINET and F. KLOPP, Enhanced Wegner and Minami estimates and eigenvalue statistics of random Anderson models at spectral edges. *Ann. Henri Poincaré* **14:5**, 1263–1285 (2013). 112, 113

[GerKlo14] F. GERMINET and F. KLOPP, Spectral statistics for random Schrödinger operators in the localized regime. *J. Europ. Math. Soc.* **16:9**, 1967–2031 (2014). 112, 113

[GreHol92] A. GREVEN and F. DEN HOLLANDER, Branching random walk in random environment: phase transitions for local and global growth rates. *Probab. Theory Relat. Fields* **91:2** 195–249 (1992). 155

[GreHol07] A. GREVEN and F. DEN HOLLANDER, Phase transitions for the long-time behaviour of interacting diffusions. *Ann. Probab.* **35:4**, 1250–1306 (2007). 170

[Gri99] G. GRIMMETT, *Percolation*, Second edition. Grundlehren der Mathematischen Wissenschaften **321**. Springer-Verlag, Berlin (1999). 92

[GrüKön09] G. GRÜNINGER and W. KÖNIG, Potential confinement property of the parabolic Anderson model. *Ann. Inst. Henri Poincaré, Probab. Stat.* **45:3**, 840–863 (2009). 52, 61, 62, 63

[GubImkPer12] M. GUBINELLI, P. IMKELLER and N. PERKOWSKI, Paraproducts, rough paths and controlled distributions. *Forum of Mathematics*, Pi **3:6** (2015). 17, 18

[GünKönSek13] O. GÜN, W. KÖNIG and O. SEKULOVIĆ, Moment asymptotics for branching random walks in random environments. *Elec. Jour. Prob.* **18**, 1–18 (2013). 155

[Hai13] M. HAIRER, Solving the KPZ equation. *Ann. Math.* **178:2**, 559–664 (2013). 17, 171

[HaiLab15a] M. HAIRER and C. LABBÉ, A simple construction of the continuum parabolic Anderson model on $\mathbb{R}^2$. *Elec. Comm. Probab.* **20**, 1–11 (2015). 17

[HaiLab15b] M. HAIRER and C. LABBÉ, Multiplicative stochastic heat equations on the whole space. preprint (2015). 18, 171

[Hal92] R.R. HALL, A quantitative isoperimetric inequality in n-dimensional space. *J. reine angew. Math.* **428**, 161–176 (1992). 125

[HarBenAvr87] S. HARVLIN and D. BEN AVRAHAM, Diffusion in disordered media. *Adv. Physics* **36**, 695–798 (1987). 1

[HofKönMör06] R. VAN DER HOFSTAD, W. KÖNIG and P. MÖRTERS, The universality classes in the parabolic Anderson model. *Commun. Math. Phys.* **267:2**, 307–353 (2006). 13, 59, 61, 62, 74, 78, 80, 88, 91, 92

[HofMörSid08] R. VAN DER HOFSTAD, P. MÖRTERS and N. SIDOROVA, Weak and almost sure limits for the parabolic Anderson model with heavy-tailed potential. *Ann. Appl. Prob.* **18**, 2450–2494 (2008). 14, 113, 116, 117

[Hol00] F. DEN HOLLANDER, *Large Deviations.* Fields Institute Monographs. American Mathematical Society (2000). 20

[HolWei94] F. DEN HOLLANDER and G.H. WEISS, Aspects of trapping in transport processes. in: G.H. Weiss (ed.), *Contemporary problems in Statistical Physics*, SIAM, Philadelphia (1994). 29

[IofVel12a] D. IOFFE and I. VELENIK, Crossing random walks and stretched polymers at weak disorder. *Ann. Probab.* **40:2**, 714–742 (2012). 145

[IofVel12b] D. IOFFE and I. VELENIK, Self-attractive random walks: the case of critical drifts. *Commun. Math. Phys.* **313:1**, 209–235 (2012). 145

[KesSid03] H. KESTEN, V. SIDORAVICIUS, Branching random walk with catalysts. *Elec. J. Prob.* **8**, 1–51 (2003). 165, 168

[KilNak07] R. KILLIP and F. NAKANO, Eigenfunction statistics in the localized Anderson model. *Ann. Henri Poincaré* **8:1**, 27–36 (2007). 112

[Kir10] W. KIRSCH, An invitation to random Schrödinger operators. *Panoramas et Syntheses* **25**, 1–119 (2010). 17, 33, 34, 36, 110

[Kom98] T. KOMOROWSKI, Brownian motion in a Poisson obstacle field. *Séminaire N. Bourbaki* exp. n⁰ 853, 91–111 (1998–9). 80

[Kön10] W. KÖNIG, Upper tails of self-intersection local times of random walks: survey of proof techniques. *Actes rencontr. CIRM* **2:1**, 15–24 (2010). 72, 142

[KönLacMörSid09] W. KÖNIG, H. LACOIN, P. MÖRTERS and N. SIDOROVA, A two cities theorem for the parabolic Anderson model. *Ann. Probab.* **37:1**, 347–392 (2009). 14, 101, 111, 116, 117, 120

[KönMuk15] W. KÖNIG and C. MUKHERJEE, Mean-field interaction of Brownian occupation measures, I: uniform tube property of the Coulomb functional. preprint (2015). 84, 144

[KönSalWol12] W. KÖNIG, M. SALVI and T. WOLFF, Large deviations for the local times of a random walk among random conductances. *Elec. Comm. Probab.* **17:10**, 1–11 (2012). 148, 149, 150

[KönSch12] W. KÖNIG and S. SCHMIDT, The parabolic Anderson model with acceleration and deceleration. In: *Probability in Complex Physical Systems*, Festschrift for Erwin Bolthausen and Jürgen Gärtner, Springer Proceedings in Mathematics Volume 11, 2012, DOI: 10.1007/978-3-642-23811-6, Heidelberg (2012). 135, 136

[KönWol15] W. KÖNIG and T. WOLFF, Large deviations for the local times of a random walk among random conductances in growing boxes. *Markov Proc. Rel. Fields* **21**, 591–638 (2015). 148, 149, 150

[LacMör12] H. LACOIN and P. MÖRTERS, A scaling limit theorem for the parabolic Anderson model with exponential potential. In: *Probability in Complex Physical Systems*, Festschrift for Erwin Bolthausen and Jürgen Gärtner, Springer Proceedings in Mathematics Volume 11, 2012, DOI: 10.1007/978-3-642-23811-6, Heidelberg (2012). 101, 117

[LieLos01] E.H. LIEB and M. LOSS. *Analysis*. 2nd Edition. AMS Graduate Studies, Vol. 14 (2001). 79

[MarRos06] M.B. MARCUS and J. ROSEN, *Markov Processes, Gaussian Processes, and Local Times*. Cambridge Univ. Press, Cambridge (2006). 82

[MerWüt01a] F. MERKL and M.V. WÜTHRICH, Annealed survival asymptotics for Brownian motion in a scaled Poissonian potential. *Stochastic Process. Appl.* **96:2**, 191–211 (2001). 82, 83, 91, 138, 143

[MerWüt01b] F. MERKL and M.V. WÜTHRICH, Phase transition of the principal Dirichlet eigenvalue in a scaled Poissonian potential. *Probab. Theory Relat. Fields* **119:4**, 475–507 (2001). 138

[MerWüt02] F. MERKL and M.V. WÜTHRICH, Infinite volume asymptotics for the ground state energy in a scaled Poissonian potential. *Ann. Inst. H. Poincaré Probab. Statist.* **38:3**, 253–284 (2002). 82, 83, 91, 138

[Mer03] F. MERKL, Quenched asymptotics of the ground state energy of random Schrödinger operators with scaled Gibbsian potentials. *Probab. Theory Relat. Fields* **126:3**, 307–338 (2003). 16, 139

[Min96] N. MINAMI, Local fluctuation of the spectrum of a multidimensional Anderson tight binding model. *Commun. Math. Phys.* **177**, 709–725 (1996). 112

[Mol81] S.A. MOLCHANOV, The local structure of the spectrum of the one-dimensional Schrödinger operator. *Commun. Math. Phys.* **78**, 429–446 (1981). 112

[Mol94] S. MOLCHANOV, Lectures on random media. In: D. Bakry, R.D. Gill, and S. Molchanov, *Lectures on Probability Theory*. Ecole d'Eté de Probabilités de Saint-Flour XXII-1992, LNM 1581, pp. 242–411. Berlin, Springer (1994). 6

[MolZha12] S. MOLCHANOV and H. ZHANG, Parabolic Anderson model with the long range basic Hamiltonian and Weibull type random potential. In: *Probability in Complex Physical Systems*, Festschrift for Erwin Bolthausen and Jürgen Gärtner, Springer Proceedings in Mathematics Volume 11, 2012, DOI: 10.1007/978-3-642-23811-6, Heidelberg (2012). 147

[Mör11] P. MÖRTERS, The parabolic Anderson model with heavy-tailed potential. In: *Surveys in Stochastic Processes*, Proceedings of the 33rd SPA Conference in Berlin, 2009. Edited by J. Blath, P. Imkeller, and S. Rœlly. EMS Series of Congress Reports. (2011). 41, 101, 116, 120, 121

[MörOrtSid11] P. MÖRTERS, M. ORTGIESE and N. SIDOROVA, Ageing in the parabolic Anderson model. *Ann. Inst. Henri Poincaré, Probab. Stat.* **47:4**, 969–100 (2011). 14, 101, 116, 120, 121

[MuiPym14] S. MUIRHEAD and R. PYMAR, Localisation in the Bouchaud-Anderson model. preprint (2014). 101, 151, 152

[MukVar15] C. MUKHERJEE and S.R.S. VARADHAN, Brownian occupation measures, compactness, and large deviations. preprint, to appear in *Ann. Probab.* (2015). 83, 84, 144

[Nak07] F. NAKANO, Distribution of localization centers in some discrete random systems. *Rev. Math. Phys* **19:9**, 941–965 (2007). 112

[Oku81] H. ÔKURA, An asymptotic property of a certain Brownian motion expectation for large time. *Proc. Japan Acad. Ser. A Math. Sci.* **57:3**, 155–159 (1981). 133

[OrtRob14] M. ORTGIESE and M. ROBERTS, Intermittency for branching random walk in heavy tailed environment. to appear in *Ann. Probab.*, preprint (2014). 156

[OrtRob16a] M. ORTGIESE and M. ROBERTS, Scaling limit and ageing for branching random walk in Pareto environment. preprint (2016). 156

[OrtRob16b] M. ORTGIESE and M. ROBERTS, One-point localization for branching random walk in Pareto environment. preprint (2016). 156

[Oss79] R. OSSERMANN, Bonessen-style isoperimetric inequalities. *Amer. Math. Monthly* **86**, 1–29 (1979). 125

[Pas77] L.A. PASTUR, The behavior of certain Wiener integrals as $t \to \infty$ and the density of states of Schrödinger equations with random potential. *Teoret. Mat. Fiz.* **32:1**, 88–95 (1977). 133

[Pov99] T. POVEL, Confinement of Brownian motion among Poissonian obstacles in R^d, $d \geq 3$. *Probab. Theory Related Fields* **114:2**, 177–205 (1999). 125

[Qua12] J. QUASTEL, Introduction to KPZ. *Current developments in mathematics*, 2011, 125–194, Int. Press, Somerville, MA (2012). 154, 171

[Res87] S.I. RESNICK, *Extreme Values, Regular Variation, and Point Processes*, Springer, New York (1987). 106, 113

[Rue14] J. RUESS, A variational formula for the Lyapunov exponent of Brownian motion in stationary ergodic potential. *ALEA* **11**, 679–709 (2014). 153, 176

[Sch26] E. SCHRÖDINGER, Quantisierung als Eigenwertproblem (Vierte Mitteilung). *Ann. Phys.* **81**, 109–139 (1926). 5

[Sch10] S. SCHMIDT, Das parabolische Anderson-Modell mit Be- und Entschleunigung (German). PhD thesis, University of Leipzig (2010). 81, 135

[Sch11] B. SCHMIDT, On a semilinear variational problem. *ESAIM Control Optim. Calc. Var.* **17**, 86–101 (2011). 64

[SchWol12] A. SCHNITZLER and T. WOLFF, Precise asymptotics for the parabolic Anderson model with a moving catalyst or trap. In: *Probability in Complex Physical Systems*, Festschrift for Erwin Bolthausen and Jürgen Gärtner, Springer Proceedings in Mathematics Volume 11, 2012, DOI: 10.1007/978-3-642-23811-6, Heidelberg (2012). 163, 164, 167

[SidTwa14] N. SIDOROVA and A. TWAROWSKI, Localisation and ageing in the parabolic Anderson model with Weibull potential. *Ann. Probab.* **42:4**, 1666–1698 (2014). 101, 118, 121, 152

[Szn91] A.-S. SZNITMAN, On the confinement property of two-dimensional Brownian motion among Poissonian obstacles. *Comm. Pure Appl. Math.* **44**, 1137–1170 (1991). 125

[Szn93] A.-S. SZNITMAN, Brownian survival among Gibbsian traps. *Ann. Probab.* **21**, 480–508 (1993). 16, 68, 95

[Szn98] A.-S. SZNITMAN, *Brownian motion, Obstacles and Random Media.* Springer-Verlag, Berlin (1998). 3, 7, 16, 23, 64, 66, 75, 80, 91, 94, 101, 105, 153

[Wag14] W. WAGNER, A random cloud model for the Schrödinger equation. *Kinet. Relat. Models* **7:2**, 361–379 (2014). 21, 177

[Wag15a] W. WAGNER, A class of probabilistic models for the Schrödinger equation. *Monte Carlo Methods Appl.* **21:2**, 121–137 (2015). 21

[Wag15] W. WAGNER, A random walk model for the Schrödinger equation. WIAS-preprint 2109, http://www.wias-berlin.de/preprint/2109/wias_preprints_2109.pdf (2015). 21

[Wol13] T.WOLFF, Random Walk Local Times, Dirichlet Energy and Effective Conductivity in the Random Conductance Model. PhD thesis, TU Berlin, http://opus4.kobv.de/opus4-tuberlin/frontdoor/index/index/docId/1490 (2013). 148, 150

[Zel84] YA.B. ZEL'DOVICH, *Selected Papers. Chemical Physics and Hydrodynamics* (in Russian). Nauka, Moscow (1984). 2

[ZelMolRuzSok87] YA.B. ZEL'DOVICH, S.A. MOLCHANOV, S.A. RUZMAJKIN and D.D. SOKOLOV, Intermittency in random media. *Sov. Phys. Uspekhi* **30:5**, 353–369 (1987). 2

[ZelMolRuzSok88] YA.B. ZEL'DOVICH, S.A. MOLCHANOV, S.A. RUZMAJKIN and D.D. SOKOLOV, Intermittency, diffusion and generation in a nonstationary random medium. *Sov. Sci. Rev. Sect. C, Math. Phys. Rev.* **7**, 1–110 (1988). 2

Index

W. König, *The Parabolic Anderson Model*, Pathways in Mathematics,
DOI 10.1007/978-3-319-33596-4

GPSR Compliance
The European Union's (EU) General Product Safety Regulation (GPSR) is a set of rules that requires consumer products to be safe and our obligations to ensure this.

If you have any concerns about our products, you can contact us on

ProductSafety@springernature.com

In case Publisher is established outside the EU, the EU authorized representative is:

Springer Nature Customer Service Center GmbH
Europaplatz 3
69115 Heidelberg, Germany

www.ingramcontent.com/pod-product-compliance
Ingram Content Group UK Ltd.
Pitfield, Milton Keynes, MK11 3LW, UK
UKHW022001270726
14060UKWH00003B/629

* 9 7 8 3 3 1 9 3 3 5 9 5 7 *